Perovskite Solar Cells

Principle, Materials and Devices

Series on Chemistry, Energy and the Environment

ISSN: 2529-7716

Series Editors: Karl M. Kadish *(University of Houston, USA)*
Roger Guilard *(University of Bourgogne, France)*

The aim of this series on "Chemistry, Energy and the Environment" is to bring together authoritative contributions where the multidisciplinary character of Chemistry and its relationship to energy and the environment can be illustrated. In each volume, the latest advances in bio-, energy- and environmentally-related fields are described in a way that unambiguously shows the major contributions from chemists at the top of their field who describe new and improved technologies for the future.

Published

Vol. 1   *Perovskite Solar Cells: Principle, Materials and Devices*
edited by Eric Wei-Guang Diau and Peter Chao-Yu Chen

Forthcoming

Vol. 2   *Elaboration and Applications of Metal-Organic Frameworks*
edited by Shengqian Ma and Jason A. Perman

Vol. 3   *Advanced Green Chemistry*
Part 1: Greener Organic Reactions and Processes
edited by István T. Horváth and Max Malacria

Vol. 4   *Prospects for Li-ion Batteries and Emerging Energy*
Electrochemical Systems, Part 1
edited by Laure Monconduit and Laurence Croguennec

Volume 1

Series on Chemistry, Energy and the Environment

Perovskite Solar Cells
Principle, Materials and Devices

Edited by

Eric Wei-Guang Diau
National Chiao Tung University, Taiwan

Peter Chao-Yu Chen
National Cheng Kung University, Taiwan

Series Editors

Karl M. Kadish
University of Houston, USA

Roger Guilard
Université de Bourgogne, France

World Scientific

NEW JERSEY · LONDON · SINGAPORE · BEIJING · SHANGHAI · HONG KONG · TAIPEI · CHENNAI · TOKYO

Published by

World Scientific Publishing Co. Pte. Ltd.

5 Toh Tuck Link, Singapore 596224

USA office: 27 Warren Street, Suite 401-402, Hackensack, NJ 07601

UK office: 57 Shelton Street, Covent Garden, London WC2H 9HE

Library of Congress Cataloging-in-Publication Data
Names: Eric Wei-Guang Diau, editor. | Peter Chao-Yu Chen
    (Professor of materials engineering), editor.
Title: Perovskite solar cells : principle, materials and devices / [edited] by
    Eric Wei-Guang Diau (National Chiao Tung University, Taiwan),
    Peter Chao-Yu Chen (National Cheng Kung University, Taiwan).
Description: New Jersey : World Scientific, 2017 | Includes bibliographical references and index.
Identifiers: LCCN 2017029412 | ISBN 9789813222519 (hardcover : alk. paper : v. 1)
Subjects: LCSH: Solar cells--Materials. | Photovoltaic cells--Materials. | Perovskite.
Classification: LCC TK2960 .P37 2017 | DDC 621.31/244--dc23
LC record available at https://lccn.loc.gov/2017029412

British Library Cataloguing-in-Publication Data
A catalogue record for this book is available from the British Library.

Typeset by Stallion Press
Email: enquiries@stallionpress.com

Printed in Singapore

Contents

3 / Sensitization and Functions of Porous Titanium Dioxide Electrodes in Dye-Sensitized Solar Cells and Organolead Halide Perovskite Solar Cells   **45**

Seigo Ito

4 / P-Type and Inorganic Hole Transporting Materials for Perovskite Solar Cells   **63**

*Ming-Hsien Li, Yu-Hsien Chiang, Po-Shen Shen,
Sean Sung-Yen Juang and Peter Chao-Yu Chen*

5 / Hole Conductor Free Organometal Halide Perovskite Solar Cells: Properties and Different Architectures

111

Sigalit Aharon and Lioz Etgar

6 / Stability Issues of Inorganic/Organic Hybrid Lead Perovskite Solar Cells

147

Dan Li and Mingkui Wang

7 / Time-Resolved Photoconductivity Measurements on Organometal Halide Perovskites   179

Eline M. Hutter, Tom J. Savenije and Carlito S. Ponseca Jr.

Preface

Due to the impact of global warming and our ever-increasing demand for energy, traditional energy sources based on fossil fuels must be replaced by clean and renewable energy. The recent Paris agreement has set a target to limit the CO_2 emission and temperature increase, for which each country shall endeavor to make contributions on solving these issues. Solar energy is considered one of the most promising, abundant and attractive sources to decrease, or even to replace, fossil fuels in the coming future. To harvest and convert solar energy effectively into electricity, various novel photovoltaic techniques have been developed. The advancement of emerging photovoltaic technology is rapid and exciting. The renaissance of organic-inorganic hybrid halide perovskite materials has particularly set off a revolutionary journey in the history of photovoltaic (and some other photonics as well) research. Their intriguing physical and chemical properties offer scientists a fantastic field to work in. Nevertheless, there are still many fundamental issues that remain to be understood and investigated. Hopefully, the success on the understanding of perovskite materials could result in practical momentum to realize commercial competition with existing fossil energy sources.

In this book, we have embraced some critical topics and expertise in different fields of perovskite solar cells research. Chapter 1 describes the important issues on the crystal growth with additive-assisted process. Chapter 2 further discusses the control of film morphology by various deposition methods. In Chapter 3, the role of porous TiO_2 for perovskite solar cells is outlined. Chapter 4 focuses on the inorganic p-type contact materials employed in perovskite solar cells. Chapter 5 reviews the properties and different architectures of hole conductor free perovskite solar cells. The critical issue of stability is the main subject of Chapter 6. Chapter 7 elucidates the characterization of perovskite solar cells using time-resolved photoconductivity technique. It is impossible to include all aspects of perovskite solar cells in this volume, for which the material development and device technology are still in rapid progress for a hope for commercialization in the near future. However, we intended to expose the readers to some of the most relevant and essential topics for recent development of perovskite solar cells in one monograph.

The content of this theme book is accomplished by the contributions from many experts in this field. We would like to express our appreciations on their great efforts and generosity during the editing process.

Eric Wei-Guang Diau, Hsinchu, Taiwan
Peter Chao-Yu Chen, Tainan, Taiwan
July 2017

1 Additive-Assisted Controllable Growth of Perovskites

Yixin Zhao* and Kai Zhu[†,‡]

*School of Environmental Science and Engineering, Shanghai Jiao Tong University, 800 Dongchuan Road, Shanghai 200240, China
[†]National Renewable Energy Laboratory, 15013 Denver West Parkway, Golden, Colorado 80401, USA
[‡]kai.zhu@nrel.gov

Table of Contents

List of Abbreviations

DSSC	dye-sensitized solar cell
EDX	energy-dispersive X-ray
FTO	fluorine-doped tin oxide
IPA	isopropanol
J_{sc}	short-circuit current density
MAI	methylammonium iodide
$MAPbI_3$	methylammonium lead iodide
PCE	power conversion efficiency
PSC	perovskite solar cells
PV	photovoltaic
SEM	scanning electron microscopy
TGA	thermogravimetric analysis
TGA-MS	thermogravimetric analysis-mass spectroscopy
V_{oc}	open circuit voltage
XPS	X-ray photoelectron spectra
XRD	X-ray diffraction

I. Introduction

Organic–inorganic hybrid perovskites (e.g., $CH_3NH_3PbI_3$ or $MAPbI_3$) were first demonstrated as a functional light absorber in the standard liquid-junction dye-sensitized solar cell (DSSC) configuration using a thick (~10 μm) mesoporous TiO_2 film by Miyasaka and coworkers in 2009.[1] In this seminal work, the facile crystallization/formation of a dark methylammonium lead iodide ($MAPbI_3$) sensitized TiO_2 layer from the precursor solution *via* simple spin coating demonstrated the promising advantages of this new semiconductor absorber compared to the complicated dye or quantum dot synthesis process. However, because of the much lower power conversion efficiency (PCE < 4%) compared to conventional DSSCs (PCE > 10%), most DSSC researchers did not focus much attention on this new absorber. Two years later, Park and coworkers significantly improved the PCE to 6.5% by using a combination of thinner mesoporous TiO_2 film, more stable liquid electrolyte, and higher-concentration precursor solution for deposition of $MAPbI_3$ perovskite.[2] This study also revealed an unusual concentration dependence of the perovskite crystallization/formation process. At that time, the crystallization process/mechanism of $MAPbI_3$ was not well understood; $MAPbI_3$ is believed to in situ crystallized on the TiO_2 surface, but interestingly, it cannot form when the precursor concentration is too low. Such observation implies a complicated crystallization of $MAPbI_3$. Although the crystallization process was not well understood, the solid-state perovskite solar cells (PSCs) soon surged in PCE to a critical value of ~10% in 2012.[3,4] This performance level almost doubled the PCE of state-of-the-art solid-state DSSCs and approached that of liquid-junction DSSCs. In these two pioneering reports, the one-step solution growth of $MAPbI_3$ from PbI_2-MAI (where MAI is methylammonium iodide) or the "mixed-halide" $MAPbI_{3-x}Cl_x$ from $PbCl_2$-3MAI exhibited a similar performance level. However, $MAPbI_{3-x}Cl_x$ soon demonstrated its success in fabricating high-efficiency PSCs with a planar structure, and also led to a long, extensive, still active debate about the role of Cl (sometimes referred to as "Cl doping fever") in lead halide perovskites. In contrast, the regular one-step method using PbI_2-MAI precursor solution seemed less ideal for producing high-quality $MAPbI_3$ thin films for PSCs. Consequently, most research groups soon switched to a two-step sequential solution deposition method to prepare high-quality $MAPbI_3$ films, in which perovskite is formed by the intercalation of MAI into PbI_2 precursor films.[5] However, the complete PbI_2-to-$MAPbI_3$ intercalation conversion *via* the two-step method is challenging in the absence of a mesoporous scaffold for providing MAI diffusion pathways.

Figure 1 shows a schematic illustration of the growth mechanism for one-step and two-step methods.[6] In the one-step method, the MA^+ and $[PbI_3]^-$ together with the coordinated solvent molecules crystallize into the perovskite structure under moderate thermal annealing. However, the formation of $MAPbI_3$ occurs simultaneously with the evaporation of solvent molecules, leading to shrinkage of the precursor films. Thus, a key factor for obtaining high-quality perovskite thin films is to control the crystallization process and solvent extraction/evaporation to avoid or mitigate the precursor film shrinkage, which is prone to create structural defects in the final perovskite films. In contrast, in the two-step method,

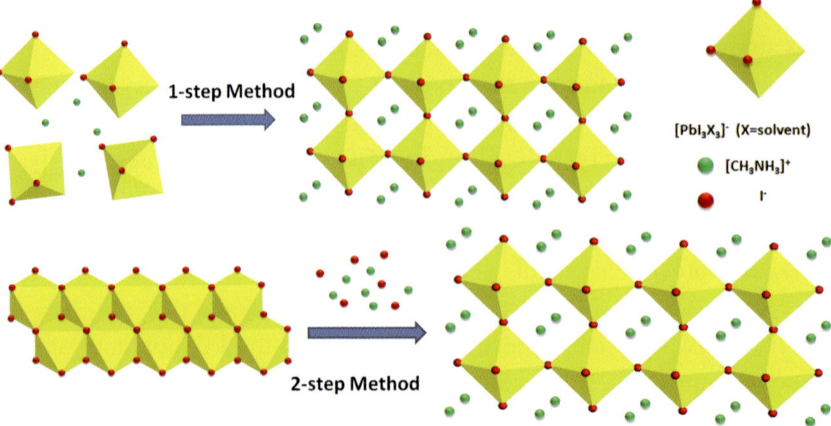

Figure 1. The schematic shrinkage and volume expansion mechanism during the crystallization of MAPbI$_3$ perovskite in one-step and two-step methods. Taken from Ref.6 with permission of the Royal Society of Chemistry.

the precursor film of PbI$_2$ undergoes substantial volume expansion (about a factor of two) after the MAI intercalation. This volume expansion can readily result in a rough surface, as opposed to the problem of volume shrinkage in the one-step method.

The additive-assisted growth methods are generally effective at controlling the shrinkage and volume expansion of precursor films in both one-step and two-step solution deposition of perovskites. In general, the additives used in one-step methods form smooth additive-containing intermediates before the final perovskite films are developed. These intermediates can effectively control perovskite crystallization kinetics and modify the film shrinkage during the perovskite film formation process. In two-step methods, the additive is usually to control the volume expansion or transformation of PbI$_2$-based precursor films into final perovskite films.

II. Cl-Based Additives

A. Role of MACl in One-Step Solution Processing

To overcome the technical obstacle to deposit high-quality perovskite films in one-step method, solution additives (especially Cl compound)

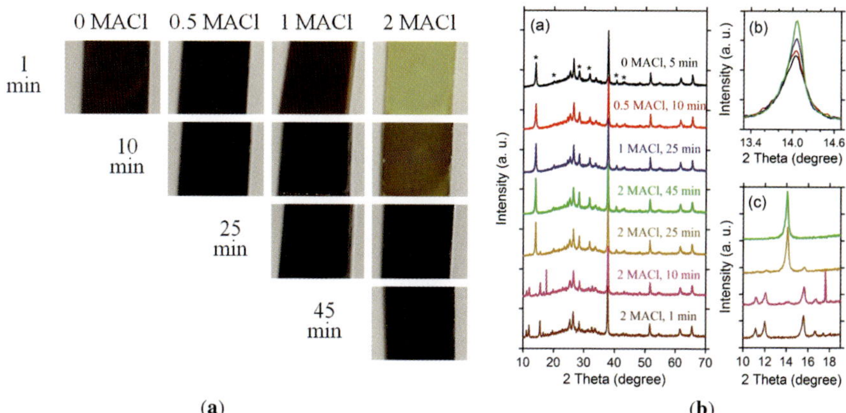

Figure 2. (a) Images illustrating the impact of different amounts of MACl and annealing duration (at 100°C) on the appearance of MAPbI$_3$ perovskite films. (b) Typical XRD patterns of the corresponding films shown in (a). Taken from Ref.7 with permission of the American Chemical Society.

have demonstrated to be effective agents to control the growth of perovskite thin films. Among various additives, the CH$_3$NH$_3$Cl or MACl, when added to the standard PbI$_2$-MAI precursor solution for one-step deposition, can effectively adjust the kinetics of the crystallization process of forming pure MAPbI$_3$ perovskite thin film, leading to enhanced crystallinity, absorption, and significantly improved coverage on a planar substrate.[7] Noteworthy is that the amount of MACl additive can vary over a wide range, enabling a large processing window for reproducible results.

Figure 2 shows the effect of varying the amount of MACl added to the one-step precursor PbI$_2$-MAI solution on the appearance of the perovskite MAPbI$_3$ films, which are annealed at 100°C for various durations as indicated. The films prepared from the regular perovskite precursor solution containing only PbI$_2$ and MAI would immediately turn to a brown film after annealing to remove the solvent. With an increasing amount of MACl, it takes more time to turn the greenish precursor perovskite film to a brown/dark brown color. For example, it takes more than 25 min for the perovskite film to turn brown when 2 molar ratio of MACl is used (hereinafter referred to as 2-MACl). The use of MACl slows down the formation of perovskite film, but it also enhances the absorbance of the final films. The color of the final perovskite film changes from brown with no MACl to dark brown with 2-MACl addition. Although the crystallization process seems different, all these perovskite films using a different amount of MACl all exhibit typical $MAPbI_3$ perovskite X-ray diffraction (XRD) patterns. The XRD evolution of $MAPbI_3$ deposited on mesoporous TiO_2 film using 2-MACl precursor shows several unknown intermediate peaks that all disappear after annealing for 45 min. This observation reveals that the addition of MACl can retard the formation of $MAPbI_3$ perovskite by forming some unknown intermediates, but the MACl additive does not affect the final crystal structure of the perovskite films. Besides the retarded crystallization, another interesting finding for the MACl additive is that these final perovskite films obtained using different amounts of MACl in the precursors do not exhibit any detectable Cl based on the elemental analysis by energy-dispersive X-ray (EDX) measurements; all samples show a final Pb:I ratio of ~1:3 after proper annealing time (depending on the amount of MACl used) to convert the precursor films to perovskite. Interestingly, when 2-MACl precursor is used, the Cl:I ratio of the precursor film is found to be ~1.6:2.7 in the absence of annealing, and it decreases to ~0.6:2.6 with 10-min annealing, further to ~0.3:2.8 with 25-min annealing, and finally becomes undetectable with 45-min annealing. These elemental analyses suggest that MACl is first incorporated into the precursor film to form some intermediates and then releases (sublimes) from the film with longer annealing time. Both the XRD and Cl content evolution suggest that MACl works as an effective additive to slow down the crystallization of $MAPbI_3$ *via* formation of Cl-incorporated intermediates.

With the retarded crystallization process, the morphology of the obtained $MAPbI_3$ perovskite planar films shows a significant dependence on the MACl amount. With the regular one-step PbI_2-MAI precursor, the $MAPbI_3$ perovskite films are usually coarse and have a needle-like morphology with low surface coverage. Such low quality would make the $MAPbI_3$ unsuitable for planar PSCs. Once MACl is added into the precursors, the perovskite films exhibit dramatic morphological changes. Figure 3 shows that when 0.5 MACl is used, the large crystal plates in the non-MACl sample disappear partially, some small crystals begin to form,

Figure 3. Scanning electron microscopy (SEM) images of MAPbI$_3$ thin films layer grown with (**a**) 0, (**b**) 0.5, (**c**) 1, and (**d**) 2 molar ratio of MACl additive in the precursor solution. Taken from Ref.7 with permission of the American Chemical Society.

and the surface coverage is enhanced. When the amount of MACl is increased, the MAPbI$_3$ film coverage enhances significantly, which is comparable to the MAPbI$_{3-x}$Cl$_x$ prepared from the mixed-halide PbCl$_2$-3MAI precursor. It is also verified that the XRD evolution and morphology of the 2-MACl sample are similar to those of the mixed-halide MAPbI$_{3-x}$Cl$_x$.[7] It is worth noting that the role of Cl (from MACl, or from PbCl$_2$, or from other sources) had been a very controversial topic for about 2 years and is still under investigation by some groups. With the enhanced morphology and crystallinity induced by the MACl additive, the photovoltaic (PV) performance of the planar MAPbI$_3$ solar cell is significantly improved to >13% compared to the 2–3% from the non-MACl case, and it is comparable to that of the mixed-halide MAPbI$_{3-x}$Cl$_x$-based planar PSCs.

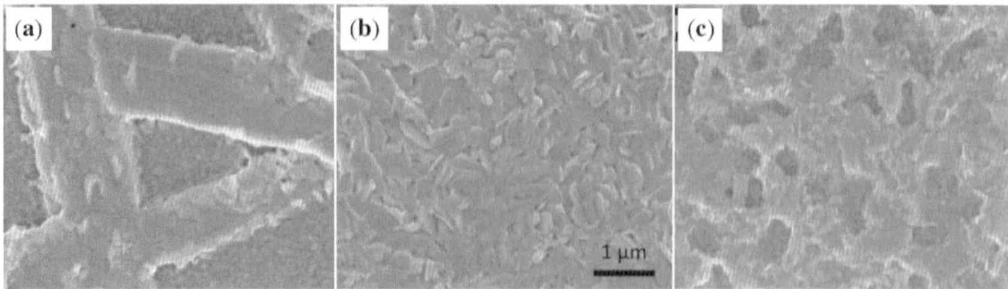

Figure 4.   Typical SEM images of MAPbI$_2$Br perovskite thin films grown from (**a**) 0, (**b**) 1, and (**c**) 2 molar ratio of MACl additives in precursor solution. Taken from Ref.8 with permission of the American Chemical Society.

The MACl additive was soon demonstrated as a general additive for MAPbX$_3$ perovskite fabrication including the 1.8-eV (band gap) mixed-halide MAPbI$_2$Br perovskite thin films for planar PSCs,[8] which could not be fabricated by the regular one-step or two-step method in early 2014. The addition of controllable MACl into PbI$_2$-MABr precursor solution helps to obtain smooth, compact MAPbI$_2$Br planar perovskite thin films without any detectable Cl, which is similar to the MACl additive–assisted growth of MAPbI$_3$. As shown in Figure 4, the MAPbI$_2$Br film prepared from precursor with only 1 molar ratio MACl (or PbI$_2$-MABr-MACl) has complete coverage on the c-TiO$_2$/FTO substrate with a compact morphology. In contrast, the PbI$_2$-MABr-2MACl precursor leads to the formation of MAPbI$_2$Br thin films with about 85% surface coverage and many interconnected small crystals. Such a difference in film morphology is ascribed to the release of different amounts of MACl in the annealing process. In some studies of the mixed-halide PbCl$_2$-3MAI system, the growth mechanism is proposed as the MAPbI$_3$-2MACl case.[9,10] However, the proposed release/sublimation of MACl during thermal annealing was also a controversial topic because pure MACl does not sublime under the normal annealing temperature (100–125°C) for MAPbI$_3$ or MAPbI$_2$Br. The thermogravimetric analysis (TGA) measurements of MACl, MABr, and MAPbI$_2$Br do not show any weight loss even when the temperature is raised to about 150–175°C (Figure 5a). The PbI$_2$·MABr·xMACl (x = 1 or 2) solid solution is proposed as a plausible intermediate film, which is investigated by the TGA measurement. The

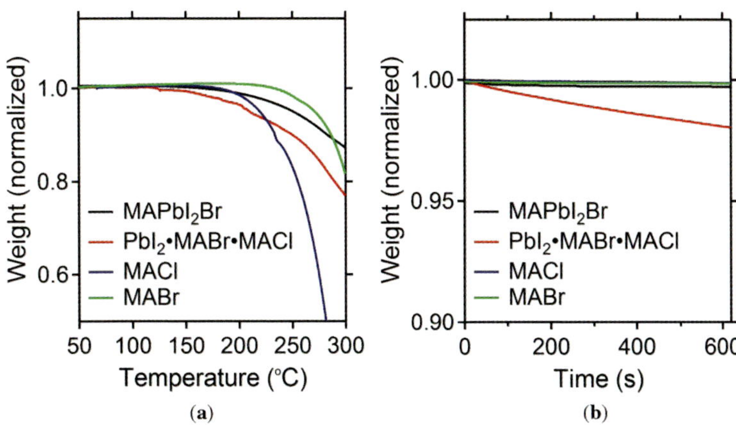

Figure 5. TGA of MACl, MABr, MAPbI$_2$Br, and PbI$_2$·MABr·MACl. (**a**) Temperature was increased from 50 to 300°C with a heating rate of 5°C/min; (**b**) temperature was held at 125°C for 600 s. Taken from Ref.8 with permission of the American Chemical Society.

starting PbI$_2$·MABr·MACl films begin to lose weight at about 125°C, and the lost weight contains Cl and Br at higher annealing temperature, which is further verified from TGA with mass spectrometry (TGA-MS). In contrast, only Cl is found in the TGA-MS measurement when the starting PbI$_2$·MABr·MACl films are kept at 125°C, suggesting that the weight loss is related to MACl sublimation.

The TGA results suggest that MACl functions like glue or a sacrificial agent to facilitate forming a compact perovskite film, and the amount of MACl could determine the final film morphology. Interestingly, the evolution of film morphology during the thermal decomposition process revealed that PbI$_2$·MABr·xMACl films are relatively compact immediately after solvent evaporation, which confirms the effect of MACl as a glue to enable the formation of compact precursor films. However, the different amount of MACl loss during the thermal annealing/decomposition from PbI$_2$·MABr·xMACl leads to a more compact MAPbI$_2$Br film using the PbI$_2$·MABr·MACl precursor than that using the PbI$_2$·MABr·2MACl precursor.

MACl has been demonstrated as a general additive to help grow high-quality perovskite (MAPbI$_3$ or MAPbI$_2$Br) thin films. MACl first

Reaction 1:

$$PbCl_2 + 3CH_3NH_3I \longrightarrow CH_3NH_3PbI_3 + 2CH_3NH_3Cl\,(g)$$

Reaction 2:

$$2\text{-}1 \quad PbCl_2 + 3CH_3NH_3I \longrightarrow PbI_2 + CH_3NH_3I + 2CH_3NH_3Cl\,(g)$$

$$2\text{-}2 \quad PbI_2 + CH_3NH_3I + 2CH_3NH_3Cl \longrightarrow CH_3NH_3PbI_3 + 2CH_3NH_3Cl\,(g)$$

$$2\text{-}3 \quad PbI_2 + xCH_3NH_3I + yCH_3NH_3Cl \longrightarrow (CH_3NH_3)_{x+y}PbI_{2+x}Cl_y(\text{Intermediate phase})$$
$$\longrightarrow CH_3NH_3PbI_3 + CH_3NH_3Cl(g)$$

Scheme 1. Two possible reaction processes related to the mixed-halide $PbCl_2$-3MAI recipe for growing $MAPbI_{3-x}Cl_x$. g represents gas phase. Taken from Ref. 9 with permission of Wiley-VCH.

alloys with the perovskite precursor to form a solid-state solution (intermediate complex) and then it released (most likely by sublimation) at a temperature lower than the decomposition temperature for perovskites. During this process, the MACl additive retards the crystallization process of lead halide perovskite to form a Cl-containing intermediate complex and then transforms to high-quality and uniform perovskite thin films without a detectable amount of Cl in the final films. MACl-assisted growth of high-quality perovskite thin films can be used to build high-efficiency planar PSCs comparable to the popular planar $MAPbI_{3-x}Cl_x$ PSCs. It is worth noting that with an improved understanding of the function and mechanism of the MACl additive, it has also been found that for the mixed-halide recipe (i.e., $PbCl_2$-3MAI) the Cl content decreases with annealing time and there is no detectable Cl in the final $MAPbI_{3-x}Cl_x$ perovskite films after the annealing. The proposed growth mechanism can be divided into two possible reaction pathways[9,10] as indicated in Scheme 1.

Based on these possible reaction pathways, the $PbCl_2$-3MAI would form $MAPbI_3$ with a byproduct of 2 (molar ratio) MACl. The MACl additive would then form an intermediate with PbI_2 and MAI, and Cl would sublime in the presence of extra MA. These results suggest that the $PbCl_2$-based precursor for preparing $MAPbI_3$ is essentially the same as the MACl-assisted methods discussed abovementioned, despite the different Cl source in these methods. The similar behavior between $PbCl_2$- and MACl-based precursors suggests that the main function of Cl additives is to retard perovskite crystallization to help form the high-quality and uniform perovskite thin films. Based on this mechanism, an alternative method using a different ratio of PbI_2:$PbCl_2$ with MAI/MABr was also demonstrated to prepare thin-film perovskites. For example, the mixed-halide perovskite of $MAPb(I_{0.8}Br_{0.2})_xCl_{3-x}$ using partial $PbCl_2$ exhibited a similar result as the MACl-additive method and exhibited similar PV performance.[11] However, it is noteworthy that the use of $PbCl_2$ can still form MACl-based additive in the solution. Thus, the separate roles of MA and Cl on the film formation process remain unclear from these pioneering studies discussed earlier. But it seems that either additional MA or Cl compound could facilitate the formation of perovskites with preferred structural properties, which will be discussed in the following sections where the development of different additives are further divided into Cl- or MA-based additives.

B. Extremely Volatile Cl-Based Additive for Room Temperature Perovskite Processing

To verify the separate roles of MA and Cl as additive to control film formation, NH_4Cl, a low-cost fertilizer, was also successfully used for preparing high-quality perovskite films.[12,13] Adding 0.5 molar ratio NH_4Cl was found to help the standard PbI_2-MAI precursor solution to produce uniform $MAPbI_3$ thin films with full surface coverage (Figure 6). Unlike

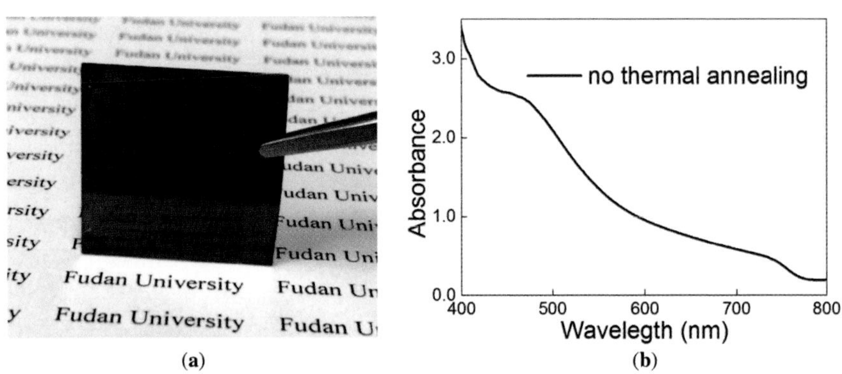

Figure 6.   (a) The images and (**b**) XRD pattern of $MAPbI_3$ prepared by using NH_4Cl additive at room temperature without using thermal annealing. Taken from Ref. 13 with permission of the American Chemical Society.

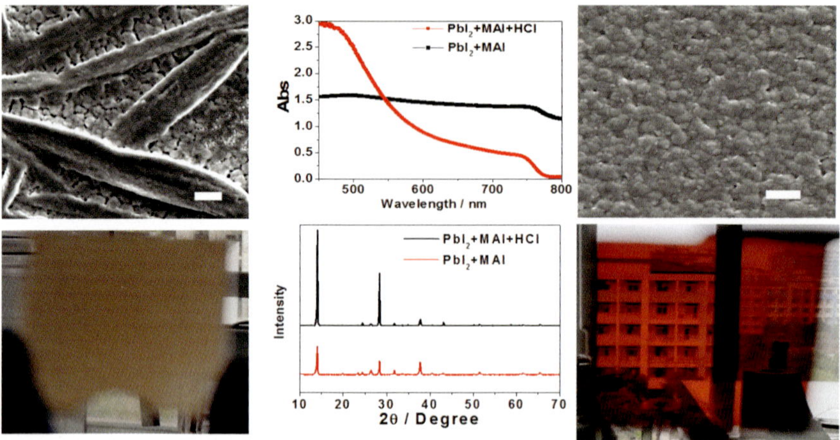

Figure 7. Comparison of SEM images, photos, UV–vis spectra, and XRD patterns of MAPbI$_3$ films deposited with and without HCl additives. The left images are the samples without HCl additive and the right images are the samples with HCl additive. Taken from Ref.14 with permission of the Royal Society of Chemistry.

most reported one-step solution deposition of perovskites, using NH$_4$Cl additive can help form highly uniform perovskite films without any thermal annealing treatment. It appears that NH$_4$Cl facilitates the crystallization of MAPbI$_3$ with rapid NH$_4$Cl sublimation during the spin-coating process.

In another study, HCl, the smallest Cl additive, is also found effective for preparing high-quality MAPbI$_3$ perovskite thin films.[14] Moreover, HCl is also found to enhance the humidity tolerance of perovskites. Under a high-humidity (~60%) environment, MAPbI$_3$ prepared with the standard PbI$_2$-MAI normally shows a brown-grey color with coarse film morphology (Figure 7, left) because the crystallization of MAPbI$_3$ is highly sensitive to the moisture environment. In contrast, adding stoichiometric HCl into PbI$_2$-MAI precursor solution leads to a very smooth planar MAPbI$_3$ film under this high-humidity level (Figure 7, right). This is attributed to the high volatile properties of HCl that accelerate MAPbI$_3$ crystallization.

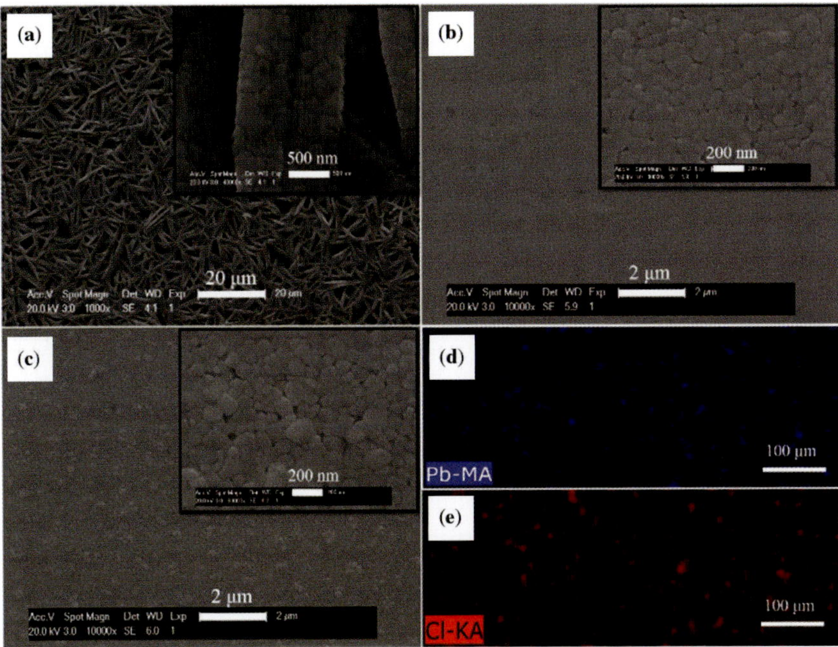

Figure 8. SEM images of (**a**) regular MAPbI$_3$ perovskite film grown without additives, (**b**) non-annealed, and (**c**) annealed perovskite films prepared with 0.5 molar ratio CaCl$_2$ additive grown at 95°C for 20 min. EDX analysis of the annealed perovskite film: (**d**) Pb map and (**e**) Cl map. Taken from Ref.15 with permission of the Royal Society of Chemistry.

C. Adverse Effect from Nonvolatile Cl-Based Additives

It is critical that Cl additives (e.g., MACl, NH$_4$Cl, or HCl) have to be volatile and easily removable in order to be effective in improving perovskite structural properties without deteriorating electrical properties. The nonvolatile Cl additives (e.g., CaCl$_2$ and NaCl) are also found to be an effective morphology controller to significantly improve the film morphology of MAPbI$_3$ perovskite.[15] But the Cl anions from CaCl$_2$ or NaCl cannot be removed as those volatile Cl additives and it appears that CaCl$_2$ or NaCl enter into the MAPbI$_3$ crystal or cover the perovskite grain boundaries (Figure 8). Consequently, the devices based on perovskites prepared from these nonvolatile Cl additives show very poor PV performance despite the improved film morphology resulting from the Cl additive. Interestingly, the deterioration of device performance is mainly due to very low short-circuit current density (J_{sc}) rather than poor open-circuit voltage (V_{oc}), suggesting that the perovskite layer is likely a mixture of insulating $CaCl_2$ and $MAPbI_3$.

III. Excess MA Additive in Nonstoichiometric Precursor

Besides the Cl additives, the MA-based additives have also become an effective approach to control perovskite morphology. The extra MA^+-additive-based method is usually realized by using a different Pb salt (such as $Pb(NO_3)_2$, $Pb(Ac)_2$, besides the common $PbCl_2$ and PbI_2) with extra MAI. Among these precursors, the $MAPbI_3$ precursor prepared by mixing MAI and $Pb(Ac)_2$ seems to be a promising candidate to obtain ultrasmooth perovskite films for high performance PSCs. The MAAc is more volatile than MACl as an additive to control the perovskite films.[16,17]

Most additives used for one-step solution deposition methods for preparing $MAPbI_3$ will go through a similar mechanism as discussed earlier for the MACl additive, as shown in Scheme 2. In general, the additive should be able to be included to form an intermediate with $MAPbI_3$; both additive and intermediate could be easily converted into perovskites, with extra materials released during the film formation process. Characteristics of a good additive may include the following: (1) it helps form a uniform precursor film; (2) it can be easily removed; (3) the additive removal process should not have a significant effect on the morphology of the intermediate film; and (4) residual additive in the perovskite film should not adversely affect the optoelectronic properties of perovskites.

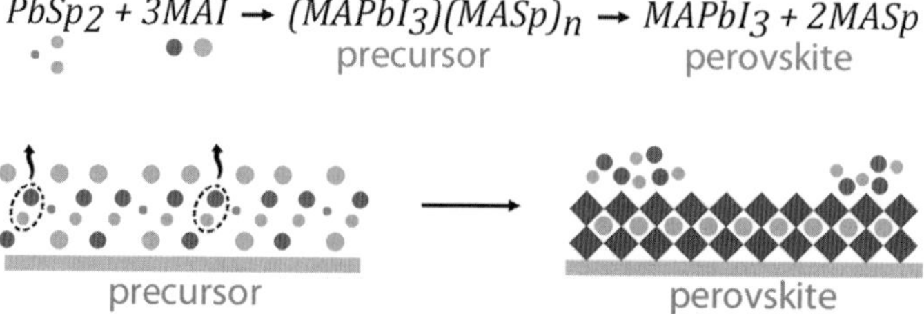

Scheme 2. The mechanism of a different additive to help form high-quality $MAPbI_3$ perovskites. Taken from Ref.17 with permission of the American Chemical Society.

Interestingly, when the amount of the organic salt additives (MAI) is significantly reduced from a molar ratio of 3 (as shown in Scheme 2), the role of MAI can be drastically enhanced. It not only facilitates the formation of compact and uniform perovskite thin films, but also improves the crystallinity and grain size and suppresses the formation of impurities in the perovskite films. A recent study demonstrates that a small amount (e.g., 20%) of excess MAI, when mixed in the standard PbI_2-MAI precursor, can promote the growth of perovskite grains up to about 1–2 μm with much enhanced crystallinity and charge-carrier lifetime.[18] It is well known that rapid crystallization (e.g., antisolvent dripping method[19]) normally results in a very compact perovskite film consisting of small grains on the order of 200–300 nm. Such films are already sufficient for achieving high-efficiency PSCs, but enlarging the grain size of the polycrystalline perovskite thin films is often beneficial for further improving the electro-optical properties of perovskites and PSC device performance. In principle, thermal annealing at an elevated temperature should be useful for grain growth. However, $MAPbI_3$ is known to be unstable under thermal annealing in air and will undergo chemical decomposition to PbI_2. Excess PbI_2 is generally not ideal for PSCs, although a small amount may be beneficial. This study shows that when excess MAI is present, it compensates for the loss of any MAI and suppresses the PbI_2 formation. This also results in the growth of small nanometer-sized perovskite grains into larger, highly textured micron-sized grains. Figure 9a shows the SEM image of a $MAPbI_3$ film prepared with the nonstoichiometric precursor (with 20% excess MAI) solution using the antisolvent bathing method.[20] Figure 9b shows the morphology of the perovskite film prepared with the same nonstoichiometric precursor and subsequently annealed at 150°C for 15 min. The grain growth is quite evident for the annealed sample where many large grains of ~1–2 μm size are seen. It is worth noting that the film thickness is only ~350 nm, and the in-plane sizes are three to six times larger. Using the nonstoichiometric precursor, the absorption spectrum for the air-annealed film is also much stronger, especially at the absorption edge, compared to the as-prepared film. Consistent with the enhanced absorption, the crystallinity of the annealed sample is also significantly enhanced (Figure 9d).

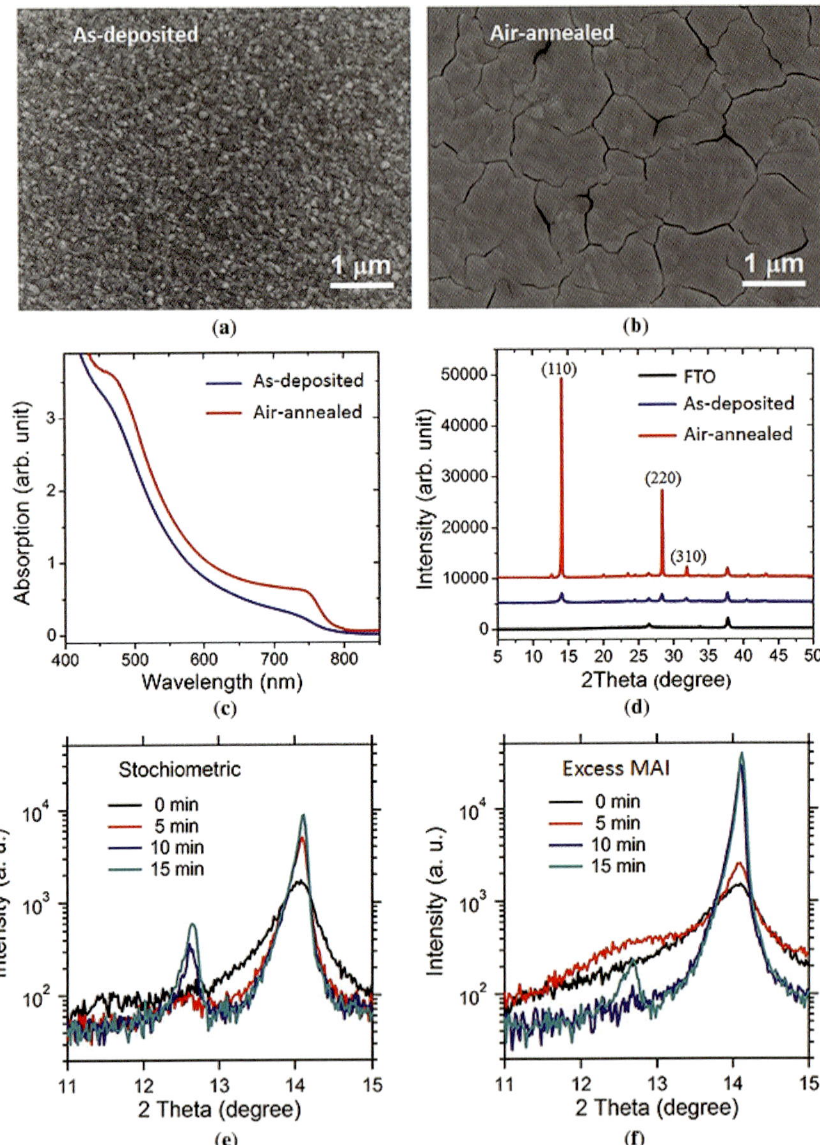

Figure 9.   SEM images (top view) of (**a**) as-deposited; (**b**) air-annealed; (**c**) absorption spectra; and (**d**) XRD patterns of as-deposited and air-annealed MAPbI$_3$ perovskite films prepared by ~20% extra MAI additives in precursor solutions; XPS of air-annealed MAPbI$_3$ perovskite films from (**e**) MAI:PbI$_2$ = 1:1, and (**f**) MAI:PbI$_2$ = 1:1.2. Taken from Ref.18 with permission of Wiley-VCH.

The key factor to the dramatic change in the structural and optical properties is the elongated time window, where perovskite undergoes growth with higher crystallinity whereas PbI_2 formation is limited. Figures 9e and 9f show the XRD patterns from both the stoichiometric- and nonstoichiometric-precursor films as a function of annealing time at 150°C. The increased intensity of the perovskite 110 peak is clear evidence of the enhanced crystallinity during annealing. In the absence of excess MAI, there is a clear indication of PbI_2 formation (~12.6°) after air annealing for 5 min. The PbI_2 peak intensity increases significantly with longer annealing time. The time window is significantly increased to around 15 min when a much weaker intensity of PbI_2 can be observed. Thus, it is clear that the excess MAI compensates for the loss of MAI, suppressing the formation of PbI_2 during air annealing. Using this non-stoichiometric $MAPbI_3$ precursor solution coupled with annealing control, the best PCE of 16.3% is achieved for a planar PSC with 1.2 cm^2 active area and the stabilized PCE output is about 15.6%.

IV. Additives for PbI_2 in Two-Step Method

The additive can help the formation of $MAPbI_3$ avoid causing significant film shrinkage in the one-step deposition method. But the additive can also be used to suppress the volume expansion of precursor films in two-step solution deposition. In 2013, Grätzel and coworkers demonstrated a 15%-efficiency PSC with a two-step sequential solution deposition method for growing perovskite thin films[5]; the two-step method was initially reported by Mitzi and coworkers in the 1990s.[21] At the early stage of PSC development, the main advantage of the two-step sequential deposition over the regular one-step method is facile realization of depositing compact and uniform perovskite film. In a typical two-step sequential solution, the perovskite film is obtained by intercalating PbI_2 with MAI, while the PbI_2 precursor films tends to form a compact, flat, layered structure. The compact and uniform PbI_2 precursor films provide an ideal template for fabrication of perovskite thin films; however, the transformation of PbI_2 into perovskite would induce volume expansion by about a factor of two. The volume expansion presents a new challenge in completely converting the dense crystalline PbI_2 film into perovskite. Because the PbI_2 residue can deteriorate the PV performance of PSCs, it is important to have complete conversion and to avoid residual PbI_2 in the two-step deposition method.[22] To complete the conversion of PbI_2 to $MAPbI_3$, Mitzi and coworkers soaked the PbI_2 film in MAI isopropanol (IPA) solution for hours.[21] Unfortunately, the long conversion time in the IPA solution of MAI can lead to the back extraction of MAI from the perovskites, further leading to partial dissolution of $MAPbI_3$. Thus, the PV performance of $MAPbI_3$ PSCs depends strongly on the soak time and often exhibits low reproducibility. To overcome this challenge, several strategies such as using elevated reaction temperature, controlling the crystallinity of the initial PbI_2 film, or using MAI vapor instead of MAI IPA solution have been developed.[23–25] Porous PbI_2 could be an ideal candidate for fast two-step deposition of perovskite films.

The first reported porous PbI_2 backbone for sequential deposition was realized by using MACl additive.[26] This porous PbI_2 film was designed and obtained *via* thermal decomposition of an unstable $MAPbI_2Cl$ film from a stoichiometric PbI_2-MACl (or PbI_2·MACl) precursor solution. The deposited $MAPbI_2Cl$ film after annealing at 130°C for 1 min exhibits a light brown color similar to $MAPbI_3$ but with a different unknown XRD pattern. This unknown compound is likely a mixture of $MAPbCl_3$, $MAPbI_3$, and some unknown structure.[27] Longer annealing time turns this brown $MAPbI_2Cl$ film to a yellow porous film with a typical PbI_2 XRD and ultraviolet–visible (UV–vis) patterns. This porous PbI_2 film formed from thermal decomposition of $MAPbI_2Cl$ turns to dark brown almost immediately (about 30 s) after dipping into the MAI solution. In this study, the MACl was used as an effective additive to fabricate a porous PbI_2 film due to the sublimation of MACl with thermal annealing.[7,8]

The twofold volume expansion associated with the intercalation processes in a typical sequential deposition of $MAPbI_3$ from PbI_2 could present several issues for preparing a compact, planar $MAPbI_3$ film: (1) MAI diffusion into the deeper layer of PbI_2 is blocked by the dense $MAPbI_3$ surface layer; and (2) volume expansion could roughen the surface morphology of perovskite. Thus, it is highly desirable to develop a novel sequential solution deposition to produce a planar $MAPbI_3$ film that has controlled morphology without any PbI_2 residue.

A facile morphology-controllable sequential deposition of planar MAPbI$_3$ film was realized by using a volume pre-expanded PbI$_2\cdot x$MAI ($x = 0.1$–0.3) precursor film with adjustable volume expansion ratio.[28] With the addition of partial MAI, the deposited PbI$_2\cdot x$MAI films still look yellowish with absorbance similar to the typical PbI$_2$ films without any characteristic absorbance associated with MAPbI$_3$. But the XRD patterns of these PbI$_2\cdot x$MAI ($x = 0$–0.3) precursor films display a weaker characteristic XRD peak at about 12.5° than the pure PbI$_2$ film. This observation suggests that a partial incorporation of MAI in the PbI$_2$ film decreases the PbI$_2$ crystallinity, which is desired for the sequential deposition. Although the addition of xMAI into PbI$_2$ does not result in any detectable MAPbI$_3$ diffraction peaks in the PbI$_2\cdot x$MAI precursors, some unknown peaks formed below 10° were observed from the PbI$_2\cdot x$MAI precursor film. It seems that the partial (<30%) intercalation of MAI into PbI$_2$ can help form certain complexes that reduce the crystallinity of the primary-phase PbI$_2$. This is similar to a previous observation of PbI$_2\cdot$DMSO-like complex formation.[19] When using PbI$_2$ as a precursor, the significant phase

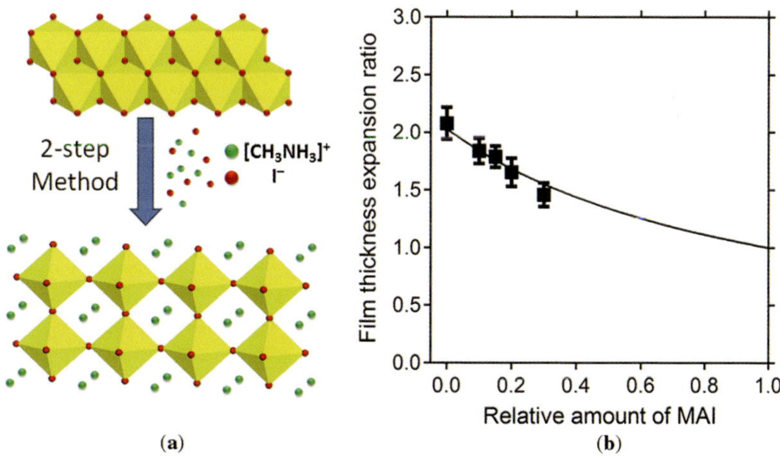

Figure 10. (a) Schematic illustration of the volume expansion from PbI$_2$ to MAPbI$_3$ in the typical two-step sequential deposition process. (b) Plot of the film thickness (or volume) expansion ratios of converting PbI$_2\cdot x$MAI to MAPbI$_3$ films as a function of the relative amount x MAI used. Taken from Ref.28 with permission of the American Chemical Society.

transformation during intercalation of MAI into PbI_2 films can lead to a twofold volume expansion; this volume expansion is verified by thickness measurement using a surface profiler. By using the additive MAI in the PbI_2 precursor film, the volume expansion from $PbI_2 \cdot xMAI$ to $MAPbI_3$ is reduced significantly; the expansion ratio decreases systematically with increasing amount of MAI. Figure 10b shows the measured volume expansion ratio as a function of the MAI amount added to the PbI_2 precursor. The result confirms that the pre-expansion of the PbI_2 films reduces the volume expansion in the sequential deposition. The plausible mechanism is that the partially added MAI in the $PbI_2 \cdot xMAI$ films is likely incorporated into the PbI_2 matrix and has partially pre-expanded the PbI_2 matrix. Assuming the pre-expansion effect or ratio of partial MAI is the same as the standard MAI intercalation process, the expansion ratio (m) of the final $MAPbI_3$ film thickness (d_f) to the thickness of the initial $PbI_2 \cdot xMAI$ film (d_i) as given by the expression $d_f/d_i = m/[1 + (m - 1)x]$ can be derived. This expression fits the data curve very well (Figure 10b) and the best fit yields a volume expansion ratio m around 2, which is consistent with the measured volume expansion ratio.

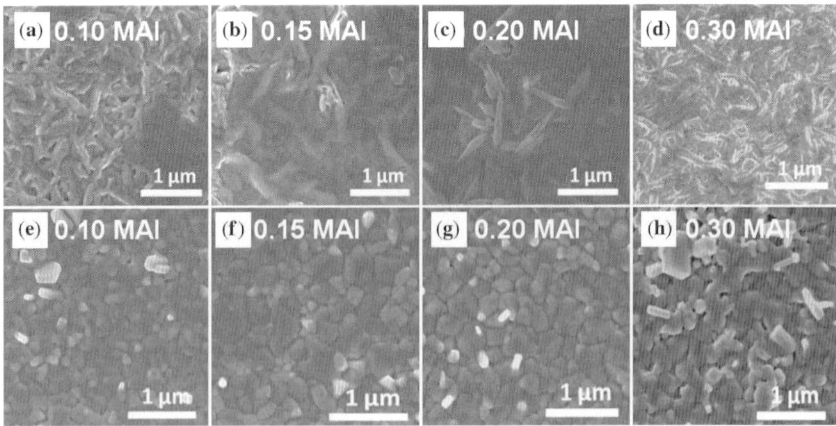

Figure 11. Typical SEM images of (**a–d**) $PbI_2 \cdot xMAI$ ($x = 0.1$, 0.15, 0.2, and 0.3) precursor films and (**e–h**) $MAPbI_3$ films prepared from their respective precursor films as indicated. Taken from Ref.28 with permission of the American Chemical Society.

After controlling the volume expansion by using $PbI_2 \cdot xMAI$ precursor films, the final $MAPbI_3$ perovskite films all become much smoother. Figure 11 shows the typical SEM images of the $PbI_2 \cdot xMAI$ ($x = 0.1$, 0.15, 0.2, and 0.3) precursor films and their final $MAPbI_3$ perovskite films after sequential deposition. First of all, these $PbI_2 \cdot xMAI$ films have very different morphologies than the pure PbI_2 film, and their morphologies vary with the amount of MAI. These $PbI_2 \cdot xMAI$ films with $x \leq 0.2$ are smoother with fewer pinholes than the pure PbI_2 film. However, the morphology of the $PbI_2 \cdot 0.3MAI$ film becomes even coarser than the pure PbI_2 film. Second, the typical SEM images of the $MAPbI_3$ films prepared from $PbI_2 \cdot xMAI$ film with $x < 0.2$ via sequential deposition are much more uniform and smoother than the $MAPbI_3$ prepared from pure PbI_2, especially when the amount of MAI is about 0.15. Furthermore, the grain sizes of the $MAPbI_3$ films also show a dependence on the amount of MAI in the $PbI_2 \cdot xMAI$ precursors. The $MAPbI_3$ film prepared from $PbI_2 \cdot 0.1MAI$ precursor mainly consists of <200-nm-sized $MAPbI_3$ nanocrystals. When using 0.15–0.2 MAI, the $MAPbI_3$ crystal size increases to about 500 nm filled with 200-nm-sized small crystals. In all, the morphology of the final $MAPbI_3$ perovskite films is comparable to that prepared from the solvent engineering or vapor-phase deposition of perovskite films. These results clearly demonstrate that pre-expanding the volume of precursor films via partial MAI addition into the standard PbI_2 precursor allows controllable two-step sequential deposition of high-quality $MAPbI_3$ film.

Consistent with the improved film morphology and phase purity of the $MAPbI_3$ prepared by two-step deposition with the $PbI_2 \cdot xMAI$ precursor, the PSC based on 0.15 MAI achieved a PCE of 17.22% with about 17% stable output and high reproducibility. This performance level was among the highest at that time based on the two-step deposition. In summary, the use of partial $xMAI$ as an additive to form a pre-expanded $PbI_2 \cdot xMAI$ precursor is a good candidate as an alternative to replace pure PbI_2 used in the two-step sequential solution deposition. This additive idea has later been adapted by other groups using a small amount of tBP or even H_2O as a good additive to form some PbI_2 x[additive].[29,30] These additives can also smooth the precursor films and help obtain high-quality perovskite thin films. Figure 12 shows an example using H_2O as the additive to control the perovskite film morphology.

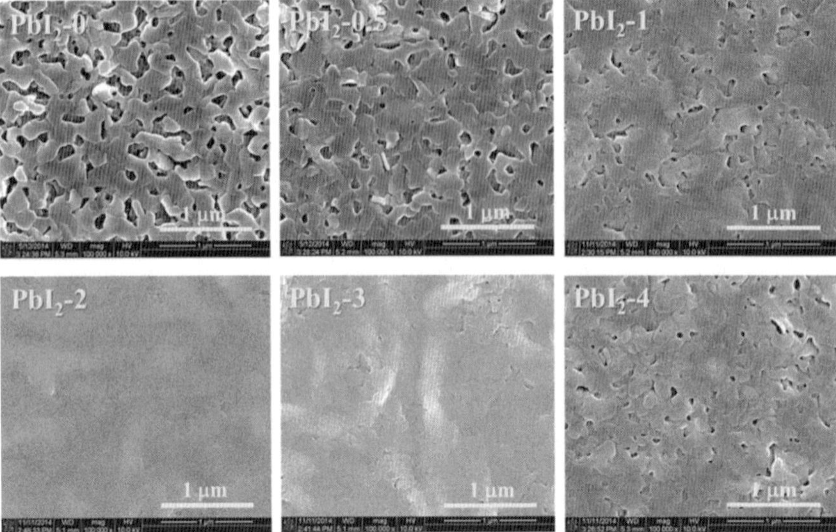

Figure 12. SEM images of PbI$_2$ films prepared from PbI$_2$/DMF solutions containing 0, 0.5, 1, 2, 3, and 4 wt % H$_2$O additive in regular PbI$_2$ precursor. Taken from Ref.29 with permission of the Royal Society of Chemistry.

To further accelerate the formation of MAPbI$_3$ *via* two-step sequential deposition, a novel precursor of HPbI$_2$Cl was developed by the facile reaction of PbI$_2$ with HCl.[14] This novel HPbI$_2$Cl precursor is different from the regular PbI$_2$ in both XRD pattern and UV–vis spectra (Figure 13a and b), but is similar to that of HPbI$_3$.[31] First, this HPbI$_2$Cl precursor can easily form a compact and uniform film as the PbI$_2$ (Figure 13c), and the precursor film obtained exhibits a high humidity tolerance. Second, the HPbI$_2$Cl precursor film does not bring any Cl residue into the final perovskite films. This is because HPbI$_2$Cl can be easily decomposed to PbI$_2$ by mild annealing at 100°C for less than 5 min, whereas HPbI$_3$ is much more stable. Unlike other additives used in the PbI$_2$ precursor — which often reduce the crystallinity of PbI$_2$ (as reflected by the reduced XRD peak) — HPbI$_2$Cl does not exhibit any PbI$_2$ characteristic peak and the crystallinity of HPbI$_2$Cl is very strong according to their XRD patterns. Interestingly, this HPbI$_2$Cl planar precursor

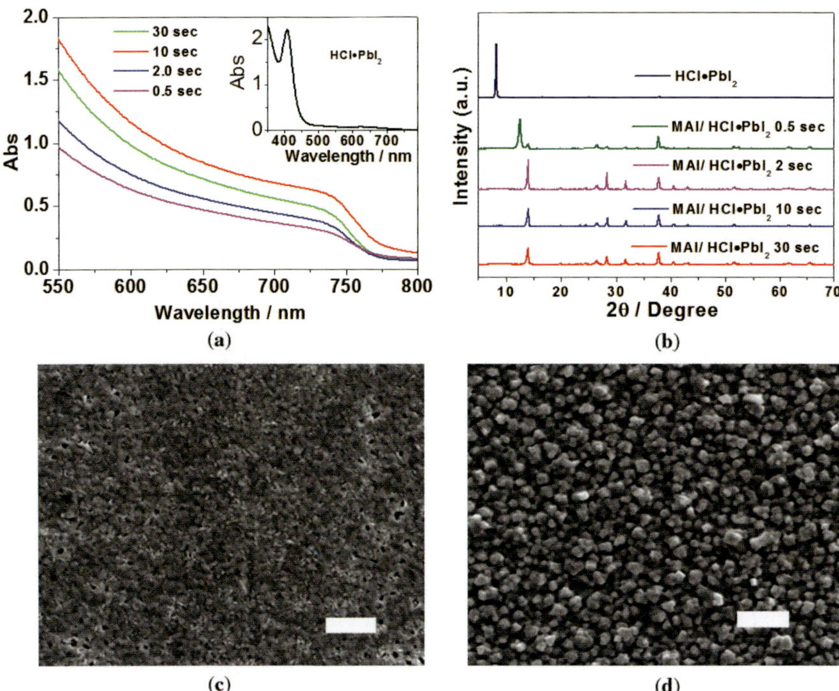

Figure 13. The (**a**) XRD patterns and (**b**) UV–vis absorption spectra evolution of the transition of planar $HCl \cdot PbI_2$ precursor film into $MAPbI_3$ by the regular two-step method; (**c**) the SEM images of $HCl \cdot PbI_2$ precursor film and (**d**) the resulting $MAPbI_3$ film. Taken from Ref.14 with permission of the Royal Society of Chemistry.

film can be completely converted into $MAPbI_3$ perovskite rapidly, within seconds, in MAI solution at room temperature. The XRD results suggest that the $HPbI_2Cl$ precursor film first converts to PbI_2 in less than 1 s and is then completely converted into $MAPbI_3$ within seconds (Figure 13b); in contrast, it usually takes tens of minutes to completely convert PbI_2 into $MAPbI_3$. Also noteworthy is that the final $MAPbI_3$ film only exhibited about 10% volume expansion compared to the $HPbI_2Cl$ precursor. It appears that the HCl additive can effectively expand the PbI_2 to $HPbI_2Cl$ with a similar volume size as $MAPbI_3$, and the unstable $HPbI_2Cl$ can easily be converted to $MAPbI_3$ *via* fast cation- and ion-exchange reaction with MAI. Due to the ultrafast sequential deposition reaction at room temperature, the $MAPbI_3$ perovskite films obtained consist of relatively small (~200 nm) crystallites and the film morphology is similar to the $MAPbI_3$ thin film prepared from pure PbI_2 use two-step sequential deposition (Figure 13d). The PSCs fabricated *via* this $HPbI_2Cl$ precursor can also exhibit ~15% efficiency, attesting to the effectiveness of this new precursor for two-step deposition.

V. Conclusions

In summary, growing high-quality perovskite thin films is one of the most important research topics for developing high-efficiency PSCs for future potential large-scale application. Understanding and controlling the crystallization kinetics and film formation process are critical for obtaining high-quality perovskite thin films by solution chemistry. Both the normal one-step and two-step solution methods have some technical obstacles for producing perovskite films for efficient devices. Fortunately, the crystallization of organic/inorganic lead halide perovskites strongly depends on the solution chemistry and can be affected by the solvent, cation/ion, and other effects. The additives — including cation/ion and some solvent molecules — can tune the growth of perovskite *via* formation of intermediates to modify the crystallization process. Generally, the additives used in the one-step methods are to form a uniform intermediate precursor film by including the additives into the perovskite precursor, and these additives can later be removed. An ideal halide additive and alkylamine additive should be easy to remove from the intermediate without leaving residues to affect the perovskite composition. Most Cl additives and CH_3NH_3 additives have been successfully adapted for preparing high-quality perovskites. Different challenges often exist in two-step solution deposition methods. To overcome the serious volume expansion in the two-step method, the first-step precursor film of mainly PbI_2 can be modified to control the volume expansion ratio by additive-assisted pre-expansion. To pre-expand these PbI_2 precursor films, the intercalation compound such as MAI can first be partially intercalated as an additive to form the $PbI_2 \cdot x$MAI precursor films. The other additives of H_2O, HI, HCl, or others can also form the intermediate precursor films with PbI_2 and can later be effectively removed during the regular second-step MAI intercalation or conversion process. Although many additives have been successfully developed for preparing high-quality perovskite thin films, the detailed mechanisms of these additives in the perovskite crystallization are still not fully understood. More fundamental studies on the additive-assisted crystallization of perovskites are necessary to further develop perovskite material growth with improved structural stability. It is believed that the additive-assisted growth of perovskites holds promise for effective solution chemistry deposition techniques and also serves as a good platform for understanding the intriguing chemical and material properties of perovskite structures and thin films.

VI. Acknowledgments

Y.Z. acknowledges the support of Natural Science Foundation of China (NSFC) (Grant 51372151 and 21303103) and Huoyingdong Grant (151046). K.Z. acknowledges the support by the hybrid perovskite solar cell program of the National Center for Photovoltaics funded by the U.S. Department of Energy, Office of Energy Efficiency and Renewable Energy, Solar Energy Technologies Office. The work at the National Renewable Energy Laboratory is supported by the U.S. Department of Energy under Contract No. DE-AC36-08-GO28308.

VIII. References

1. Kojima, A.; Teshima, K.; Shirai, Y.; Miyasaka, T. *J. Am. Chem. Soc.* **2009**, *131*, 6050–6051.

2. Im, J.-H.; Lee, C.-R.; Lee, J.-W.; Park, S.-W.; Park, N.-G. *Nanoscale* **2011**, *3*, 4088–4093.

3. Chung, I.; Lee, B.; He, J.; Chang, R. P. H.; Kanatzidis, M. G. *Nature* **2012**, *485*, 486–489.

4. Kim, H.-S.; Lee, C.-R.; Im, J.-H.; Lee, K.-B.; Moehl, T.; Marchioro, A.; Moon, S.-J.; Humphry-Baker, R.; Yum, J.-H.; Moser, J. E.; Grätzel, M.; Park, N.-G. *Sci. Rep.* **2012**, *2*, 591.

5. Burschka, J.; Pellet, N.; Moon, S.-J.; Humphry-Baker, R.; Gao, P.; Nazeeruddin, M. K.; Grätzel, M. *Nature* **2013**, *499*, 316–319.

6. Zhao, Y.; Zhu, K. *Chem. Soc. Rev.* **2016**, *45*, 655–689.

7. Zhao, Y.; Zhu, K. *J. Phys. Chem. C*, **2014**, *118*, 9412–9418.

8.   Zhao, Y.; Zhu, K. *J. Am. Chem. Soc.* **2014**, *136*, 12241–12244.

9.   Yu, H.; Wang, F.; Xie, F.; Li, W.; Chen, J.; Zhao, N. *Adv. Funct. Mater.* **2014**, *24*, 7102–7108.

10.   Dualeh, A.; Tétreault, N.; Moehl, T.; Gao, P.; Nazeeruddin, M. K.; Grätzel, M. *Adv. Funct. Mater.* **2014**, *24*, 3250–3258.

11.   Liang, P.-W.; Chueh, C.-C.; Xin, X.-K.; Zuo, F.; Williams, S. T.; Liao, C.-Y.; Jen, A. K. Y. *Adv. Energy Mater.* **2015**, *5*, 1400960.

12.   Rong, Y.; Hou, X.; Hu, Y.; Mei, A.; Liu, L.; Wang, P.; Han, H. *Nature Commun.* **2017**, *8*, 14555.

13.   Chen, Y.; Zhao, Y.; Liang, Z. *Chem. Mater.* **2015**, *27*, 1448–1551.

14.   Li, G.; Zhang, T.; Zhao, Y. *J. Mater. Chem. A* **2015**, *3*, 19674–19678.

15.   Chen, Y.; Zhao, Y.; Liang, Z. *J. Mater. Chem. A* **2015**, *3*, 9137–9140.

16.   Zhang, W.; Saliba, M.; Moore, D. T.; Pathak, S. K.; Hörantner, M. T.; Stergiopoulos, T.; Stranks, S. D.; Eperon, G. E.; Alexander-Webber, J. A.; Abate, A.; Sadhanala, A.; Yao, S.; Chen, Y.; Friend, R. H.; Estroff, L. A.; Wiesner, U.; Snaith, H. J. *Nat. Commun.* **2015**, *6*, 6142.

17.   Moore, D. T.; Sai, H.; Tan, K. W.; Smilgies, D.-M.; Zhang, W.; Snaith, H. J.; Wiesner, U.; Estroff, L. A. *J. Am. Chem. Soc.* **2015**, *137*, 2350–2358.

18.   Yang, M.; Zhou, Y.; Zeng, Y.; Jiang, C.-S.; Padture, N. P.; Zhu, K. *Adv. Mater.* **2015**, *27*, 6363–6370.

19.   Jeon, N. J.; Noh, J. H.; Kim, Y. C.; Yang, W. S.; Ryu, S.; Seok, S. *Nat. Mater.* **2014**, *13*, 897–903.

20.   Zhou, Y.; Yang, M.; Wu, W.; Vasiliev, A. L.; Zhu, K.; Padture, N. P. *J. Mater. Chem. A* **2015**, *3*, 8178–8184.

21.   Liang, K.; Mitzi, D. B.; Prikas, M. T. *Chem. Mater.* **1998**, *10*, 403–411.

22.   Zhao, Y.; Nardes, A. M.; Zhu, K. *Farad. Discuss.* **2014**, *176*, 301–312.

23.   Wu, Y.; Islam, A.; Yang, X.; Qin, C.; Liu, J.; Zhang, K.; Peng, W.; Han, L. *Energy Environ. Sci.* **2014**, *7*, 2934–2938.

24.   Docampo, P.; Hanusch, F. C.; Stranks, S. D.; Döblinger, M.; Feckl, J. M.; Ehrensperger, M.; Minar, N. K.; Johnston, M. B.; Snaith, H. J.; Bein, T. *Adv. Energy Mater.* **2014**, *4*, 1400355.

25.   Lou, Y.; Zhao, Y.; Chen, J.; Zhu, J.-J. *J. Mater. Chem. C* **2014**, *2*, 595–613.

26.   Zhao, Y.; Zhu, K. *J. Mater. Chem. A* **2015**, *3*, 9086–9091.

27.   Park, B.-w.; Philippe, B.; Gustafsson, T.; Sveinbjörnsson, K.; Hagfeldt, A.; Johansson, E. M. J.; Boschloo, G. *Chem. Mater.* **2014**, *26*, 4466–4471.

28.   Zhang, T.; Yang, M.; Zhao, Y.; Zhu, K. *Nano Lett.* **2015**, *15*, 3959–3963.

29.   Wu, C.-G.; Chiang, C.-H.; Tseng, Z.-L.; Nazeeruddin, M. K.; Hagfeldt, A.; Grätzel, M. *Energy Environ. Sci.* **2015**, *8*, 2725–2733.

30.   Zhang, H.; Mao, J.; He, H.; Zhang, D.; Zhu, H. L.; Xie, F.; Wong, K. S.; Grätzel, M.; Choy, W. C. H. *Adv. Energy Mater.* **2015**, *5*, 1501354.

31.   Wang, F.; Yu, H.; Xu, H.; Zhao, N. *Adv. Funct. Mater.* **2015**, *25*, 1120–1126.

2 Control of Film Morphology for High-Performance Perovskite Solar Cells

Cheng-Min Tsai, Hau-Shiang Shiu, Hui-Ping Wu and Eric Wei-Guang Diau*

Department of Applied Chemistry and Institute of
Molecular Science, National Chiao Tung University,
Hsinchu, Taiwan
*diau@mail.nctu.edu.tw

Table of Contents

List of Abbreviations

AS   antisolvent
DSSC   dye-sensitized solar cell
EEL   electron-extraction layer
ETL   electron-transport layer
HTM   hole-transporting material
HEL   hole-extraction layer
HTL   hole-transport layer
MAI   methylammonium iodide
PCE   power conversion efficiency
PSC   perovskite solar cells
SA   solvent annealing
TA   thermal annealing

I. Introduction

Organometallic halide perovskites are promising photovoltaic materials because the rapid development of their device performance attained a power conversion efficiency (PCE) exceeding 20%,[1] comparable to the performance of an Si-based solar cell. Perovskite can serve as both light absorber and carrier transporter in a thin layer[2–4]; the corresponding devices are constructed like a dye-sensitized solar cell (DSSC) with a mesoscopic structure,[5–8] or like an organic solar cell with a planar hetero-junction configuration.[9–12] The perovskite solar cells (PSCs) can be fabri-cated according to either a one-step[6,13] or a sequential[7,8] method to synthesize the required perovskite layer on top of the contact electrode. To obtain an extraordinary device performance, the kinetics of the nucleation and crystal growth of perovskite must, however, be carefully controlled so that a well-organized perovskite film free of pinholes and with large grains can be produced. For this purpose, various synthetic strategies of solvent engineering,[14–18] self-induced passivation,[19] and gas-assisted crystallization[20] are reported to generate uniform perovskite crystals for highly efficient *n*-type or *p*-type PSC. Alternatively, the crystallinity and film morphology of perovskite can be controlled with additives of var-ied types using a one-step spin-coating method to improve the device performance.[12,21–25]

For the mesoscopic n-type PSCs, a dense TiO_2 compact layer (c-TiO_2) of thickness ~20 nm was deposited on etched fluorine-doped tin oxide (FTO) substrates by spray pyrolysis with titanium diisopropoxide bis(acetylacetonate) (75 mass % in isopropanol, 0.125 M; Aldrich) solution at 450 °C. The mesoporous TiO_2 layer (m-TiO_2) of thickness ~200 nm was then made with HD1-type TiO_2 nanocrystals (particle size ~30 nm) with spin coating on the c-TiO_2/FTO substrate at 6000 revolutions per minute (rpm) for 15 s and sintering at 500 °C for 30 min. The perovskite layer was made using an antisolvent (AS) treatment with thermal annealing (TA) or with solvent annealing (SA). For the AS + TA approach, the perovskite precursor solution [45% in dimethylformamide (DMF)] was dripped onto the substrate spinning at a rate 5000 rpm for a total spin period 30 s; during the spinning, chlorobenzene (anhydrous, 99.8%; Aldrich) serving as an AS was injected onto the substrate at a delay from 2 to 8 s. The substrates were then annealed at 100 °C for 10 min. For the AS + SA approach, DMF solvent (about 10 µL) was added on a hot plate at 100 °C to form DMF vapor; at the same time, a glass Petri dish covered both the AS-treated substrate and the DMF droplet for SA in 10 min. A hole-transport layer (HTL) was deposited with a solution containing Spiro-OMeTAD (82.5 mg; Luminescence Technology) in chlorobenzene (1 mL), Li-bis(trifluoro-methanesulfonyl)imide (Li-TFSI, 7.8 mg; Aldrich) in acetonitrile (15.6 µL) and 4-*tert*-butylpyridine (TBP, 96%, 22.5 µL; Aldrich) spinning at 2200 rpm for 30 s.

For the planar p-type PSCs, the etched ITO substrates were cleaned and spun with a commercial solution (PEDOT:PSS, AI4083; UR) at 5000 rpm for 50 s, then sintered at 120 °C for 10 min. For the deposition of the perovskite layer the methods (AS + TA and AS + SA) were the same as those applied in making n-type devices. To form the electron-extraction layer (EEL), [6,6]-phenyl- C_{61}-butanoic acid methyl ester (PCBM, 20 mg; FEM Tech) was dissolved in chlorobenzene (1 mL) and coated on the top of the perovskite layer at spinning rate 1000 rpm for 30 s. To complete the fabrication for both n- and p-type devices, thermal evaporation was applied to deposit the silver back-contact electrode. All fabrication was undertaken in a glove box under N_2.

In lack of a scaffold layer in a PSC with a planar configuration, the AS method is a powerful tool to synthesize a uniform and dense perovskite film. According to this approach, an AS is injected after a critical delay when the spin-coated perovskite solution attains a condition of supersaturation before crystal growth. Homogeneous crystallization thus results after an appropriate annealing treatment. For example, Cheng and coworkers[14] and Seok and coworkers[15] ultilized this method followed by TA at 100 °C to prepare *n*-type PSC with PCE exceeding 16%. Huang and coworkers developed a gentle procedure called SA, in which a solvent vapor activated the reaction between methylammonium iodide (MAI) and lead iodide (PbI_2) layers to form a uniform perovskite layer for a *p*-type planar PSC with PCE 15.6%.[17] In the recent work, perovskite nanocrystals was synthesized with the AS approach in combination with either TA or SA to fabricate an *n*-type mesoscopic or a *p*-type planar heterojuction solar cell. The AS + SA method can generate uniform perovskite nanocrystals close packed in a polycrystalline single layer with large grains; this approach produces *p*-type planar PSC free of hysteresis with device performance PCE 14.4%, which is improved by more than 50% relative to that of the traditional AS + TA approach. In contrast, the *n*-type mesoscopic device fabricated using the AS + SA method had serious interfacial defects. As the first example, the enhanced performance of the *n*-type mesoscopic PSC fabricated with the AS + TA method and oxidation of the *p*-doped HTL overnight before evaporation of the silver back contact was demonstrated. As a result, remarkable photovoltaic performance was obtained with PCE 16.5% with little effect of hysteresis. The crystallization kinetics significantly control the film morphology to generate uniform perovskite crystals by accelerating nucleation and decelerating crystal growth *via* SA.

II. Solvent-Assisted Crystallization and Film Morphology

Figure 1 shows a synthetic approach to produce uniform and well-packed perovskite nanocrystals with the AS method followed by SA. For the AS approach, the perovskite precursor solution (45% in DMF, PbI_2 + MAI in equal molar ratio) was dripped onto a spinning substrate at 25 °C. After a set delay, a second solvent (chlorobenzene) was quickly injected onto the

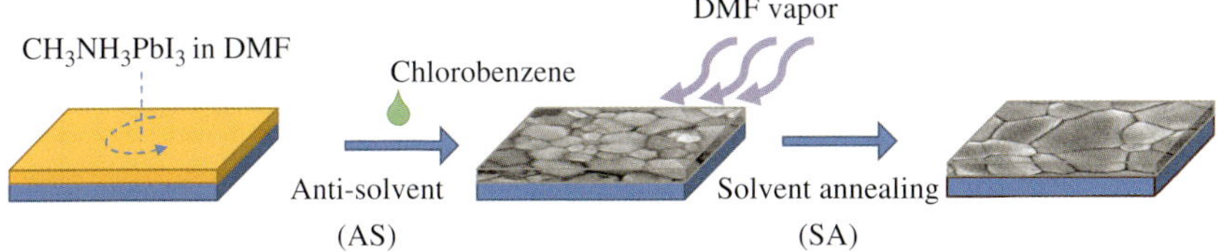

Figure 1. Synthesis of uniform perovskite nanocrystals *via* an AS treatment with chlorobenzene followed by SA with DMF vapor.

Figure 2. Top-view SEM images showing morphological changes of perovskite nanocrystals prepared with AS treatment with delays (**a**) 4, (**b**) 6, and (**c**) 8 s followed by TA for 10 min, and prepared *via* the AS treatment with delay 6 s followed by SA for (**d**) 5, (**e**) 10, and (**f**) 30 min.

substrate; the total spinning procedure lasted 30 s. The crystal growth proceeded when the AS-treated films were subjected to a traditional TA at 100 °C for 10 min. The AS method utilizes chlorobenzene as an AS; it forms a homogeneous perovskite solution with DMF but cannot dissolve perovskite crystals, to attain supersaturation of the solution within a specific delay. In the top-view scanning electron microscopy (SEM) images shown in Figure 2, the dense perovskite films free of pinholes were formed only at a delay of 6 s (Figure 2b), a smaller delay of 4 s (Figure 2a), and a greater delay of 8 s (Figure 2c), leading to the formation of irregular

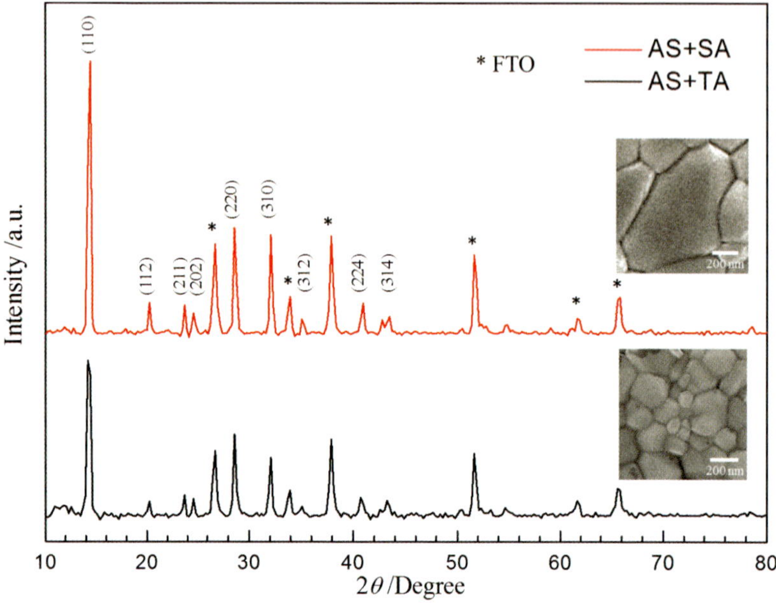

Figure 3. XRD patterns of perovskite films fabricated according to a AS approach following SA (AS + SA, top trace) and with AS approach following TA (AS + TA, bottom trace). Diffraction signals of the FTO substrate are marked with*.

perovskite crystals with uncovered surfaces, such as those reported by Xiao *et al.*[14] based on a planar *n*-type film configuration.

For the AS + SA approach, instead of TA, SA with the AS-treated films (delay 6 s) was used under DMF vapor for 10 min. Figure 2d–f shows top-view SEM images for periods 5, 10, and 30 min of SA, respectively. Comparing the results between AS + TA and AS + SA, SA helps to increase the size of crystals and further improves the film uniformity and crystallinity [X-ray diffraction (XRD) results, Figure 3]. Specifically, the crystal size increased from ~300 nm to ~1 μm for SA increased from 5 to 10 min, but no appreciable change in crystal morphology was found for SA from 10 to 30 min. Hence, the perovskite nanocrystals were synthesized according to the conditions of SA (delay 6 s) + TA (10 min) and SA (delay 6 s) + SA (10 min) to optimize the performance of the corresponding PSC devices.

III. Fabrication of *N*- and *P*-Type Perovskite Solar Cells

Both *n*- and *p*-type PSCs were fabricated according to the device configurations shown in Figure 4a and b, respectively; the corresponding potential-energy diagrams appear in Figure 4c and d, respectively. For *n*-type devices, a mesoscopic structure with a thin c-TiO$_2$ and a m-TiO$_2$ served as electron-transport layer (ETL); spiro-OMeTAD served as the HTL. For *p*-type devices, a planar heterojunction structure was applied with PEDOT:PSS and PCBM served as hole-extraction layer (HEL) and EEL, respectively. For devices of both types, a silver layer was thermally evaporated to serve as a back-contact electrode.

Figure 5a and b shows side-view SEM images for *n*- and *p*-type devices, respectively, fabricated with the AS + TA method; Figure 4c and d

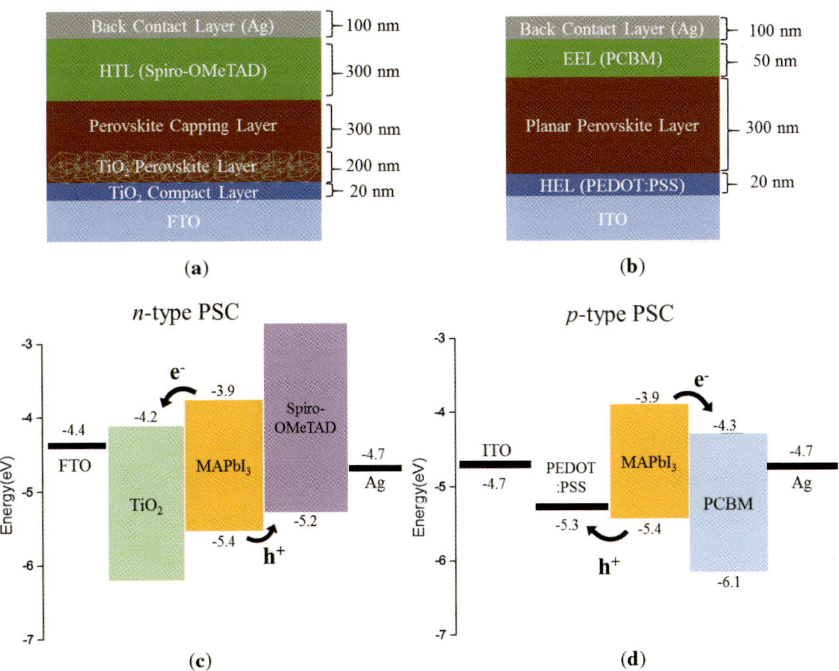

Figure 4.   Film configurations for (**a**) *n*-type and (**b**) *p*-type PSCs with corresponding potential-energy diagrams shown in (**c**) and (**d**), respectively.

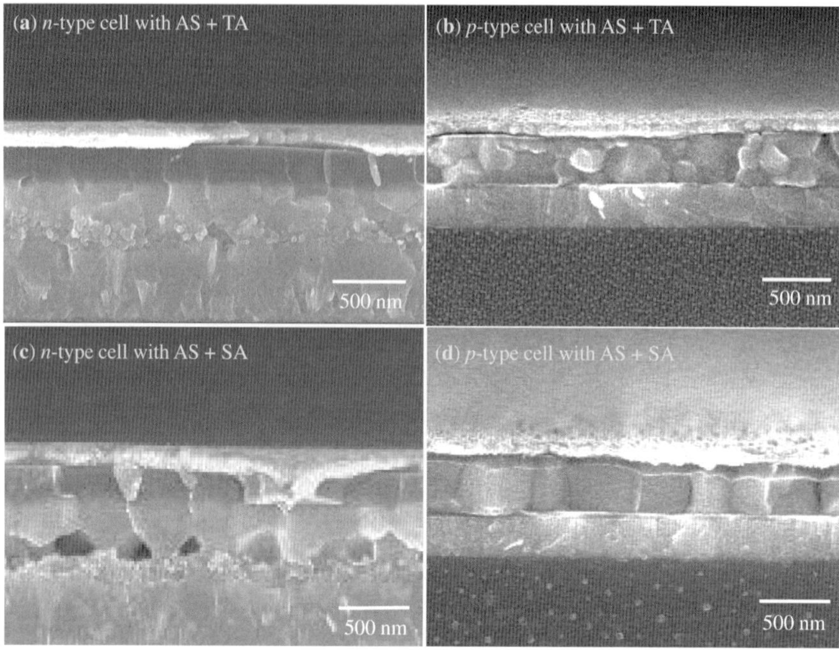

Figure 5. Side-view SEM images showing film configurations of PSCs fabricated with an AS treatment followed by TA for (**a**) *n*-type and (**b**) *p*-type devices and with AS followed by SA for (**c**) *n*-type and (**d**) *p*-type devices.

show those images for *n*- and *p*-type devices, respectively, fabricated with the AS + SA method. For *n*-type mesoscopic devices, a smooth and uniform perovskite film with a capping layer of thickness ~300 nm was observed after TA (Figure 5a), but bubble holes existed in the m-TiO$_2$/perovskite interface with the SA method (Figure 5c). The SA method evidently worked poorly with a mesoporous substrate because annealing for crystal growth occurred from both heating the substrate at the top surface through the DMF vapor and at the bottom side *via* direct thermal transfer. As a result, defects were generated at the TiO$_2$/perovskite interface as shown in Figure 5c for films produced with the AS + SA method. The results thus indicate that the mesoscopic *n*-type PSC would be expected to work better with the AS + TA method than with the AS + SA method.

For *p*-type planar devices from the AS + SA approach, both the grain size and crystallinity of the perovskite crystals are significantly improved. As shown in Figure 5b, with the AS + TA method perovskite crystals of grain size ~100 nm accumulated to form a perovskite layer of thickness ~300 nm, but the grain size of the crystals expanded to ~300 nm with close-packed nanocrystals in a single-layer pattern (Figure 5d). Huang and coworkers reported close-packed perovskite crystals produced *via* an interdiffusion method with separate PbI_2 and MAI layers post-treated with the same SA procedure.[17] These results (AS + SA) are similar to those of Huang and coworkers (interdiffusion + SA) but with superior crystal uniformity. A perovskite layer containing well-packed nanocrystals is an important feature for efficient charge transport inside the solar cells because no grain boundaries would exist along the charge-transport (vertical) direction.

IV. Photovoltaic Performance of *N*-Type and *P*-Type Devices

The morphological features shown in Figure 4 reflect the corresponding photovoltaic performances. Figure 6a and b show the current–voltage characteristics for *n*- and *p*-type devices, respectively. In each plot, the *J–V* curves for the devices with perovskite layers generated *via* the AS + TA and the AS + SA methods are shown together for comparison; the corresponding photovoltaic parameters are listed in Table 1. For mesoscopic *n*-type PSC devices with configuration $FTO/c\text{-}TiO_2/m\text{-}TiO_2/perovskite/$ spiro-OMeTAD/Ag, the device fabricated with the AS + TA method exhibited photovoltaic performance with PCE 15.5%, which is superior to that fabricated with the AS + SA method (PCE 10.9%). These results are expected because mesoscopic devices made with SA generated interfacial defects as shown in Figure 4c. The mesoscopic *n*-type device fabricated according to the AS + TA approach thus exhibited photovoltaic performance superior to that of Grätzel and coworkers (PCE = 15.0%)[7] with a classic two-step sequential method to produce the perovskite layer. Park, Grätzel, and coworkers used also a two-step spin-coating method to synthesize large perovskite crystals with cuboid morphology for *n*-type mesoscopic PSC with a device performance attaining PCE 17.0%,[8] but the one-step synthetic approach is simpler than that with a controlled crystal

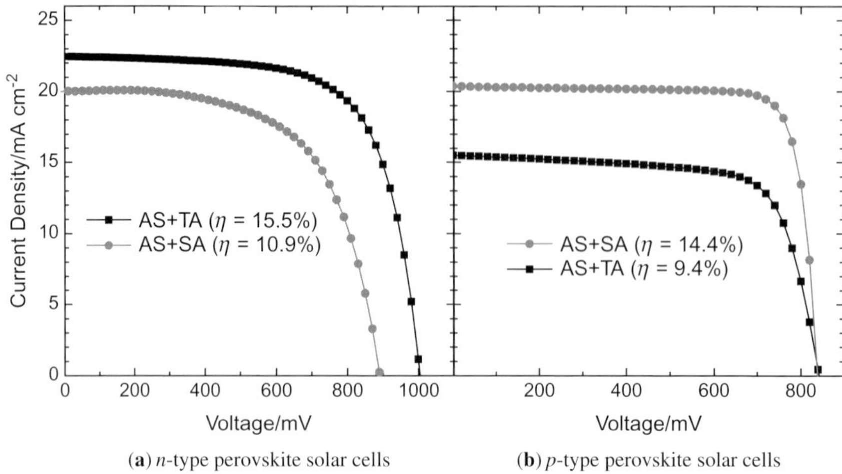

(**a**) *n*-type perovskite solar cells   (**b**) *p*-type perovskite solar cells

Figure 6. Current-voltage characteristics with an AS treatment followed by TA and with AS followed by SA for (**a**) *n*-type and (**b**) *p*-type devices.

Table 1. Photovoltaic parameters of *n*- and *p*-type devices with perovskite nanocrystals fabricated with AS and TAAS + TA and AS followed by SA (AS + SA) under simulated AM-1.5G illumination (power density 100 mW·cm^{-2}) with active areas 0.09 cm^2 and 0.04 cm^2 for *n*- and *p*-type devices, respectively.

Device configuration	Perovskite film	J_{SC}/mA·cm^{-2}	V_{OC}/mV	FF	η/%
n-type (mesoscopic)	AS + TA	22.46	1010	0.685	15.5
	AS + SA	20.02	892	0.613	10.9
p-type (planar)	AS + TA	15.48	843	0.718	9.4
	AS + SA	20.39	845	0.834	14.4

FF, fill factor.

morphology to form uniformly packed and smooth perovskite films free of pinholes.

In contrast, planar *p*-type PSC devices with configuration ITO/PEDOT:PSS/perovskite/PCBM/Ag benefit from uniform close-packed perovskite crystals *via* SA, for which the PCE of the device significantly improved from 9.4% *via* the AS + TA method to 14.4% *via* the AS + SA method. Twenty identical devices were fabricated under the same experimental conditions for the mesoscopic *n*-type and planar *p*-type devices

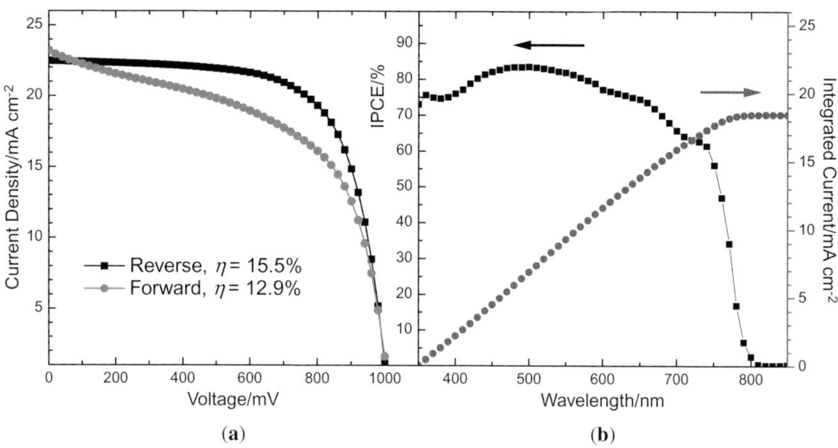

Figure 7.   Photovoltaic performance for n-i-p mesoscopic PCS fabricated using the AS + TA method showing (**a**) current–voltage characteristics *via* scans from reverse (from open-circuit to short-circuit condition, squares) and forward (from short-circuit to open-circuit condition, circles) directions and (**b**) the corresponding IPCE spectrum with integrated current density (18.4 mA·cm^{-2}, circles).

using the AS + TA and the AS + SA methods, respectively. The average photovoltaic parameters of these 20 mesoscopic n-type devices are J_{SC}/mA·cm^{-2} = 21.6 ± 0.8, V_{OC}/mV = 1016 ± 24, fill factor (FF) = 0.66 ± 0.01, and PCE/% = 14.4 ± 0.7; those of the planar p-type devices are J_{SC}/mA·cm^{-2} = 19.5 ± 0.8, V_{OC}/mV = 853 ± 28, FF = 0.825 ± 0.01, and PCE/% = 13.7 ± 0.3. The J–V curves and the corresponding incident photon to current efficiency (IPCE) spectra of the best performing n-type device are shown in Figure 7a and b, respectively; those of the best performing p-type device are shown in Figure 8a and b, respectively.

V. Effect of Hysteresis and Oxidation of Hole-Transport Materials

The effect of hysteresis is also tested for the best performing n- and p-type devices with J–V characteristic scans in both reverse (from open-circuit to short-circuit) and forward (from short-circuit to open-circuit) directions; the corresponding J–V scan results are shown in Figures 7a and 8a, respectively. The planar p-type device displays essentially no hysteresis

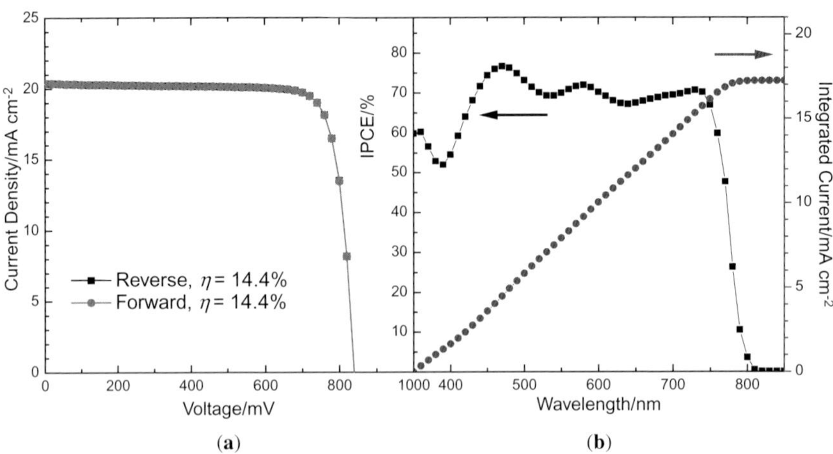

Figure 8. Photovoltaic performance for p-i-n planar PCS fabricated using the AS + SA method showing (**a**) current–voltage characteristics *via* scans from reverse (from open-circuit to short-circuit condition, squares) and forward (from short-circuit to open-circuit condition, circles) directions and (**b**) the corresponding IPCE spectrum (squares) with integrated current density (17.2 mA·cm⁻², circles).

effect whereas the mesoscopic n-type device shows a serious effect of hysteresis. The hysteresis-free feature of the p-type planar device is consistent with the results reported by Im and coworkers[12] with the same device configuration. For the mesoscopic n-type device, serious J–V hysteresis might be due to charge accumulation either in trap states of the perovskite/TiO$_2$ interface or in the HTL (spiro-OMeTAD) involving Li-TFSI and TBP additives as p-dopants. Snaith and coworkers[26,27] mentioned that oxidation of spiro-OMeTAD by Li-TFSI and oxygen plays a key role in the doping mechanism to decrease charge transport resistance and to improve the conductivity of HTL for solid-state DSSCs. This idea is applied here to examine the effect of J–V hysteresis for mesoscopic n-type PSC with p-doped spiro-OMeTAD as the hole-transporting materials (HTMs).

Using p-doped PTAA instead of p-doped spiro-OMeTAD as an HTL, Seok and coworkers[15] reported a mesoscopic n-type device with a negligible effect of hysteresis: PCE 15.9 and 15.8% in reverse and forward scans, respectively. Because the spiro-OMeTAD layer (thickness 300 nm) is much thicker than the PTAA layer (thickness 50 nm), it is suspected that

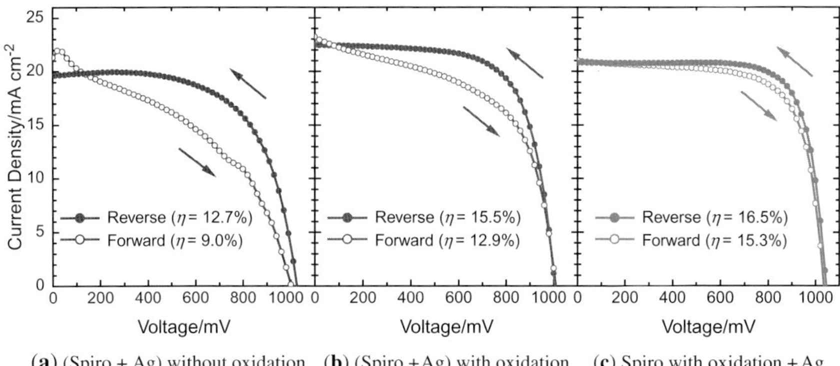

(**a**) (Spiro + Ag) without oxidation   (**b**) (Spiro + Ag) with oxidation   (**c**) Spiro with oxidation + Ag

Figure 9.   Reverse and forward scans for current–voltage curves of *n*-type devices with the AS + TA treatment showing the effect of hysteresis under varied spiro oxidation conditions: (**a**) without oxidation; (**b**) with oxidation overnight after thermal evaporation of an Ag back contact; (**c**) with overnight oxidation before thermal evaporation of Ag back contact.

the effect of hysteresis might be due to the retardation of hole transport inside the devices. Hence, the photovoltaic performance of the *n*-type devices was examined with overnight oxidation (~18 h) of the *p*-doped spiro-OMeTAD layer under three conditions: (1) without overnight oxidation, (2) with overnight oxidation after evaporation of Ag, and (3) with overnight oxidation before evaporation of Ag. The current–voltage curves of these three devices are shown in Figure 9a to c, respectively, and the corresponding photovoltaic parameters are summarized at Table 2.

As a reference for condition (1), the device performance without overnight oxidation is poor, with serious *J–V* hysteresis reflecting a poor FF value for the forward scan. Condition (2) is a normal oxidation procedure for the complete device stored in an air-filled dry box (temperature 25 °C, humidity 50%) for 18 h before the *J–V* measurements. Although the device performance in condition (2) improved significantly, oxidation of the entire device did not rectify the problem of hysteresis through the protection of the outer Ag layer. To resolve this problem, in condition (3), the device is put into a dry box for oxidation overnight before evaporation of Ag. The photovoltaic results follow: for the forward scan, J_{SC}/mA·cm^{-2} = 21.0, V_{OC}/mV = 1040, FF = 0.699, and PCE = 15.3%; for the reverse scan, J_{SC}/mA·cm^{-2} = 20.8, V_{OC}/mV = 1050, FF = 0.755, and PCE = 16.5%.

Table 2. Photovoltaic parameters of *n*-type PSCs without overnight oxidation (without ox), with overnight oxidation after thermal evaporation of a silver back-contact layer (ox after Ag), and with overnight oxidation before thermal evaporation of a silver back-contact layer (ox before Ag) under simulated AM-1.5G illumination (power density 100 mW·cm^{-2}) with active areas 0.09 cm^2.

Overnight oxidation (ox)	Scan direction	J_{SC}/mA·cm^{-2}	V_{OC}/mV	FF	η/%
(1) without ox	reverse	19.61	1030	0.627	12.7
	forward	22.62	995	0.398	9.0
(2) ox after Ag	reverse	22.46	1010	0.685	15.5
	forward	23.22	1010	0.550	12.9
(3) ox before Ag	reverse	20.80	1050	0.755	16.5
	forward	21.01	1040	0.699	15.3

With the device oxidation according to condition (3), a photovoltaic performance was obtained comparable to that of Seok and coworkers, PCE = 16.7%,[15] using *p*-doped PTAA as HTL. Hence, it is demonstrated that this prior-oxidation procedure is effective to improve the FF and overall photovoltaic performance for *n*-type mesoscopic PSCs using *p*-doped spiro-OMeTAD as the HTL.

The concept of formation of a uniform perovskite layer from a solution is to control the kinetics of crystallization so that homogeneous nucleation occurs before the crystal to further grow up. As demonstrated in Figure 10a for the AS approach, a chlorobenzene droplet was placed onto the substrate with a varied delay during which the spin-coated perovskite solution reached supersaturation. Chlorobenzene here served as an AS to accelerate the nucleation so that a dense and uniform film became produced *via* homogeneous crystallization from the following TA treatment. As demonstrated in Figure 10b for the SA approach, crystal growth was slowed when annealing was assisted with DMF vapor.

VI. Conclusion

Combining AS with the SA method shows that the grains of perovskite nanocrystals can be significantly enlarged with a uniform close-packed

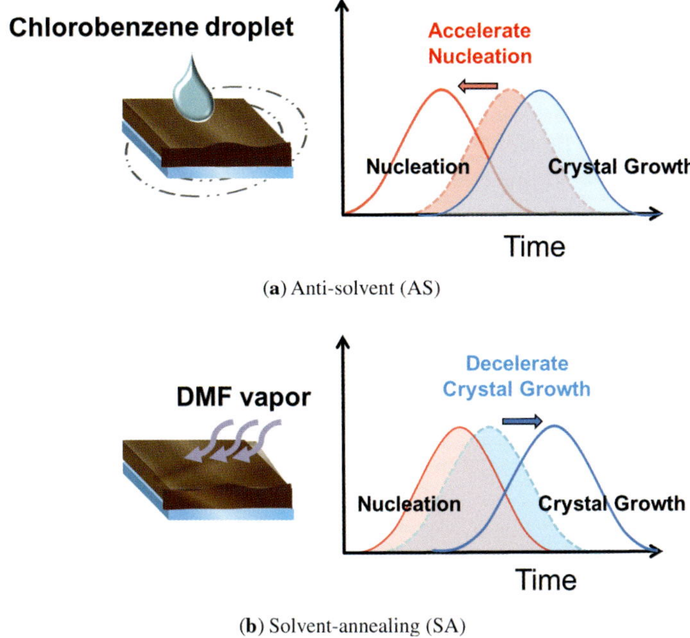

(a) Anti-solvent (AS)

(b) Solvent-annealing (SA)

Figure 10. Schematic illustration of the concept of (**a**) AS and (**b**) SA treatments.

morphology in a single layer pattern. This morphological feature is beneficial for a planar device for which the performance attained PCE 14.4% without any hysteresis for a p-type planar heterojunction PSC fabricated with the AS + SA method. This AS + SA approach worked poorly, however, with a mesoscopic device because defects were involved at the interface between perovskite and m-TiO$_2$ layers. After prior oxidation for the p-doped spiro-OMeTAD layer, the photovoltaic performance of an n-type mesoscopic PSC using the AS + TA method became optimized to give PCE 16.5% with little effect of hysteresis. The AS + SA method is hence a powerful tool to control the film morphology featuring excellent uniformity and crystallinity for the fabrication of a highly efficient planar type PSC. For n-type devices with p-doped HTM such as spiro-OMeTAD, overnight oxidation should be performed before evaporation of the silver back-contact electrode.

VII. Acknowledgments

The authors thank Professor Tzung-Fang Guo and Professor Peter Chen of National Cheng Kung University (NCKU) (Taiwan) for their assistance on *p*-type device fabrications. The initio studies done by Mr. Chien-Yi Chan and Mr. Wei-Ting Chou are acknowledged. The funding was supported by Ministry of Science and Technology (MOST) of Taiwan and National Chiao Tung University (NCTU) of Taiwan.

VIII. References

1. Yang, W. S.; Noh, J. H.; Jeon, N. J.; Kim, Y. C.; Ryu, S.; Seo, J.; Seok, S. I. *Science* **2015**, *348,* 1234–1237.
2. Rhee, J. H.; Chung, C.-C.; Diau, E. W.-G. *NPG Asia Mater.* **2013**, *5,* e68.
3. Gao, P.; Grätzel, M.; Nazeeruddin, M. K. *Energy Environ. Sci.* **2014**, *7,* 2448–2463.
4. Zhou, Z.; Pang, S.; Liu, Z.; Xu, H.; Cui, G. *J. Mater. Chem. A.* **2015**, *3,* 19205–19217.
5. Kojima, A.; Teshima, K.; Shirai, Y.; Miyasaka, T. *J. Am. Chem. Soc.* **2009**, *131,* 6050–6051.
6. Kim, H.-S.; Lee, C.-R.; Im, J.-H.; Lee, K.-B.; Moehl, T.; Marchioro, A.; Moon, S.-J.; Humphry-Baker, R.; Yum, J.-H.; Moser, J. E.; Grätzel, M.; Park, N.-G. *Sci. Rep.* **2012**, *2,* 591.
7. Burschka, J.; Pellet, N.; Moon, S.-J.; Humphry-Baker, R.; Gao, P.; Nazeeruddin, M. K.; Grätzel, M. *Nature* **2013**, *499,* 316–319.
8. Im, J.-H.; Jang, I.-H.; Pellet, N.; Grätzel, M.; Park, N.-G. *Nature Nanotech.* **2014**, *9,* 927–932.
9. Liu, M.; Johnston, M. B.; Snaith, H. J. *Nature* **2013**, *501,* 395–398.
10. Zhou, H.; Chen, Q.; Li, G.; Luo, S.; Song, T.-b.; Duan, H.-S.; Hong, Z.; You, J.; Liu, Y.; Yang, Y. *Science* **2014**, *345,* 542–546.
11. Zhang, L. Q.; Zhang, X. W.; Yin, Z. G.; Jiang, Q.; Liu, X.; Meng, J. H.; Zhao, Y. J.; Wang, H. L. *J. Mater. Chem. A* **2015**, *3,* 12133–12138.
12. Heo, J. H.; Han, H. J.; Kim, D.; Ahn, T. K.; Im, S. H. *Energy Environ. Sci.* **2015**, *8,* 1602–1608.
13. Lee, M. M.; Teuscher, J.; Miyasaka, T.; Murakami, T. N.; Snaith, H. J. *Science* **2012**, *338,* 643–647.
14. Xiao, M.; Huang, F.; Huang, W.; Dkhissi, Y.; Zhu, Y.; Etheridge, J.; Gray-Weale, A.; Bach, U.; Cheng, Y. B.; Spiccia, L. *Angew. Chem. Int. Ed.* **2014**, *53,* 9898–9903.
15. Jeon, N. J.; Noh, J. H.; Kim, Y. C.; Yang, W. S.; Ryu, S.; Seok, S. I. *Nature Mater.* **2014**, *13,* 897–903.
16. Eperon, G. E.; Burlakov, V. M.; Docampo, P.; Goriely, A.; Snaith, H. J. *Adv. Funct. Mater.* **2014**, *24,* 151–157.

17. Xiao, Z.; Dong, Q.; Bi, C.; Shao, Y.; Yuan, Y.; Huang, J. *Adv. Funct. Mater.* **2014**, *26*, 6503–6509.

18. Zhou, Y.; Yang, M.; Wu, W.; Vasiliev, A. L.; Zhu, K.; Padture, N. P. *J. Mater. Chem. A.* **2015**, *3*, 8178–8184.

19. Chen, Q.; Zhou, H.; Song, T.-B.; Luo, S.; Hong, Z.; Duan, H.-S.; Dou, L.; Liu, Y.; Yang, Y. *Nano Lett.* **2014**, *14*, 4158–4163.

20. Huang, F.; Dkhissi, Y.; Huang, W.; Xiao, M.; Benesperi, I.; Rubanov, S.; Zhu, Y.; Lin, X.; Jiang, L.; Zhou, Y.; Gray-Weale, A.; Etheridge, J.; McNeill, C. R.; Caruso, R. A.; Bach, U.; Spiccia, L.; Cheng, Y.-B. *Nano Energy* **2014**, *10*, 10–18.

21. Liang, P.-W.; Liao, C.-Y.; Chueh, C.-C.; Zuo, F.; Williams, S. T.; Xin, X.-K.; Lin, J.; Jen, A. K. Y. *Adv. Mater.* **2014**, *26*, 3748–3754.

22. Eperon, G. E.; Stranks, S. D.; Menelaou, C.; Johnston, M. B.; Herz, L. M.; Snaith, H. J. *Energy Environ. Sci.* **2014**, *7*, 982–988.

23. Heo, J. H.; Song, D. H.; Im, S. H. *Adv. Mater.* **2014**, *26*, 8179–8183.

24. Zuo, C.; Ding, L. *Nanoscale* **2014**, *6*, 9935–9938.

25. Wang, F.; Yu, H.; Xu, H.; Zhao, N. *Adv. Funct. Mater.* **2015**, *25*, 1120–1126.

26. Abate, A.; Hollman, D. J.; Teuscher, J.; Pathak, S.; Avolio, R.; D'Errico, G.; Vitiello, G.; Fantacci, S.; Snaith, H. J. *J. Am. Chem. Soc.* **2013**, *135*, 13538–13548.

27. Abate, A.; Leijtens, T.; Pathak, S.; Teuscher, J.; Avolio, R.; Errico, M. E.; Kirkpatrik, J.; Ball, J. M.; Docampo, P.; McPherson, I.; Snaith, H. J. *Phys. Chem. Chem. Phys.* **2013**, *15*, 2572–2579.

3 Sensitization and Functions of Porous Titanium Dioxide Electrodes in Dye-Sensitized Solar Cells and Organolead Halide Perovskite Solar Cells

Seigo Ito*

Department of Materials and Synchrotron Radiation Engineering,
Graduate School of Engineering, University of Hyogo, 2167 Shosha,
Himeji, Hyogo 671-2280, Japan
*ito6ne@gmail.com

Table of Contents

List of Abbreviations

FF	fill factor
FTO	fluorine doped tin oxide
HOMO	highest occupied molecular orbital
HTM	hole-transporting material
IPCE	incident photon-to-current conversion efficiency
J_{sc}	short-circuit current density
LUMO	lowest unoccupied molecular orbital
PCE	photoenergy conversion efficiency
V_{oc}	open-circuit voltage
XRD	X-ray diffraction

I. History of Dye Sensitization and Dye-Sensitized Solar Cells

Before dye sensitization, the surface photoelectrochemical reaction on semiconductor was discussed by Fujishima and Honda using TiO_2 (titanium dioxide) photoelectrode in aqueous electrolyte (named as the "Honda–Fujishima effect"), which produced hydrogen and oxygen by irradiation of sunlight.[1] The reaction scheme is shown in Figure 1 and categorized as follows:

(1) Light absorption by electron in TiO_2 valence band and electron excitation to conduction band

(2) Electron transfer in TiO_2 conduction band to back-contact electrode

(3) Hole transfer to the surface of TiO_2 to react with electrolyte in solution

(4) Electron transfer in external circuit.

(5) Reduction of electrolyte in solution

(6) Diffusion of electrolyte to transfer charges

(7) Reduction of TiO_2/oxidation of electrolyte

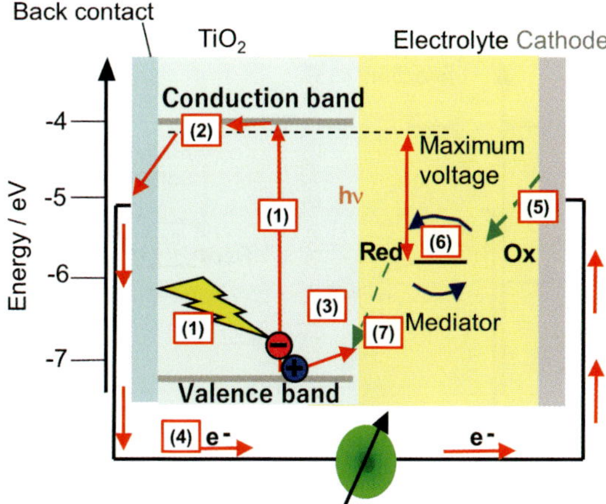

Figure 1. Photocatalysis by TiO$_2$ (the Honda–Fujishima effect).

Under the continuous irradiation, each step can occur simultaneously. At first, it was thought that the future energy problem for human beings would be solved by the Honda–Fujishima effect. However, TiO$_2$ can be active under ultraviolet (UV) light because of the wide band gap (3.0–3.2 eV). Because UV light (<400 nm) in the sunlight was only 4%, the Honda–Fujishima effect had not been able to become a new energy source.

Afterward, the enhancement of light absorption range to wider wavelength (narrower band gap) has been considered by "sensitization." Although TiO$_2$ is transparent in visible light region due to the wide band gap, a dye adsorbed on TiO$_2$ can absorb visible light and can inject electron into the TiO$_2$ conduction band, resulting in charge separation using visible light (Figure 2). The reaction scheme is as follows:

(1) Light absorption by electron in highest occupied molecular orbital (HOMO) level and electron excitation to lowest occupied molecular orbital (LUMO) level of the dye

(2) Electron transfer from LUMO level of the dye to TiO$_2$ conduction band

(3) Electron transfer in TiO$_2$ conduction band from the surface of nanocrystalline-TiO$_2$ porous electrode to the back contact (conducting glass)

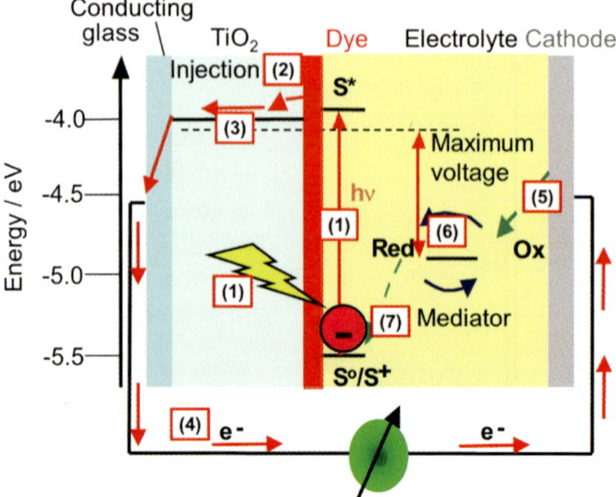

Figure 2.   Scheme of dye sensitization.

(4) Electron transfer in external circuit
(5) Reduction of electrolyte in solution
(6) Diffusion of redox electrolyte to transfer negative and positive charges
(7) Reduction of dye/oxidation of electrolyte

The significant differences of the Honda–Fujishima effect and dye sensitization are as follows:

(1) The places of electron excitation and separation are different (TiO_2 crystal bulk for the Honda–Fujishima effect; TiO_2 surface for dye sensitization).
(2) The hole transfer in TiO_2 valence band is necessary for the Honda–Fujishima effect.
(3) Oxidation energy of electrolyte was different due to the difference of TiO_2 valence band and HOMO level of the dye. The hole in TiO_2 valence band can oxidize H_2O to O_2.

The photoresponse of TiO_2 electrode in visible range can be improved by adsorption of dye on TiO_2 surface (Figure 3). This improvement has been called "dye sensitization."

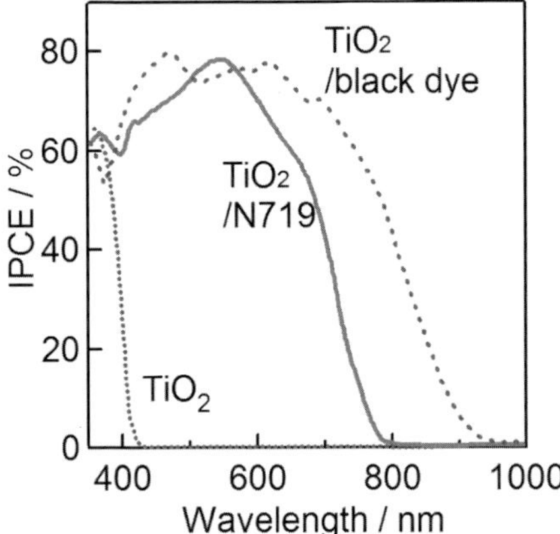

Figure 3. Incident photon-to-current conversion efficiencies (IPCEs) obtained with Ru dyes (N719 and black dye). Dye can improve the light absorption range to be "dye sensitization." Adapted from Ref. 2.

The history of dye sensitization can be considered with investigation of photograph. The first photoactive material was reported by Johann Heinrich Schulze in 1719. He found chalk–silver mixture changing from white to black with light irradiation and concluded that silver chloride (AgCl) has photosensitization.[3] Hundred years later, in 1825, Joseph Nicéphore Niépce took the world's first picture as "a man leading a horse." He used AgCl at first, but it could not work properly. In place of AgCl, he succeeded taking picture using "Bitumen of Judea," a naturally occurring asphalt.[4] Afterward, Frederick Scott Archer investigated collodion process for taking photographs using $AgNO_3$ and iodide in 1851. And then, Richard Leach Maddox investigated photographic plate using KBr and $AgNO_3$ dried with gelatin on glass plates in 1871.[5] However, because the band gaps of AgBr and AgCl are 2.6 and 3.0 eV, photosensitizations of AgBr and AgCl can be active with only blue or UV light (480 and 400 nm), respectively. Afterward, Hermann Wilhelm Vogel found dye sensitization (using aniline dyes) to enhance the visible light activity of silver halides in 1873, which is the basis of present film-type photograph.[6]

About electrochemistry of dye sensitization, James Moser reported electron transportation from dye to semiconductor as photosensitization in 1887.[7] After the finding of the Honda–Fujishima effect[1] and necessity of the visible light activation, dye sensitization has been researched actively.

Tsubomura and Mutsumura (Osaka University, Japan) and coworkers reported dye sensitization on porous ZnO electrode.[8] Their findings were as follows:

(1) ZnO was micrometer sized.
(2) Dye did not get adsorbed on the ZnO surface but dissolved in electrolyte solution.
(3) Efficiency of monochromatic light of wavelength 563 nm was 2.5%.

Afterward, Michael Grätzel [École Polytechnique Fédérale de Lausanne (EPFL) Switzerland] improved the efficiency of dye-sensitized solar cells (also known as Grätzel cells) as follows[9]:

(1) Nanocrystalline-TiO_2 electrodes were used for dye sensitization.
(2) Sensitizing dyes were adsorbed on semiconductor electrodes.
(3) A photoenergy conversion efficiency (PCE) of 7% was obtained under 1 sun conditions (AM 1.5G, 100 mW·cm^{-2}) for dye-sensitized solar cells.

Sensitization using nanocrystalline-TiO_2 electrode can enhance the photocurrent close to 1000 times than using single-crystal TiO_2 one, due to the 1000-times-larger surface area.[10] Since then, the function of nanocrystalline-TiO_2 electrode has been studied actively.

After the investigation of organolead halide perovskite solar cells,[11,12] which emerged as a family of dye-sensitized solar cells, still nanocrystalline-TiO_2 electrode has been used for the highest-efficiency perovskite solar cells.[13,14]

II. Functions of Porous TiO$_2$ Electrode

In this review, following three significant points of nanocrystalline-TiO_2 electrode for dye-sensitized solar cells and organolead halide perovskite solar cells were introduced:

(1) Tuning of particle and hole size (–> light transparency and material diffusion)
(2) Electron transportation and accumulation at the TiO_2 surface
(3) Prohibition of short circuit between working and counter electrodes

Each point is quite important to understand the system of dye-sensitized and perovskite solar cells.

III. Tuning of Particle and Hole Size (Light Transparency and Material Diffusion)

Size of TiO_2 nanoparticle is one of the important factors of perovskite and dye-sensitized solar cells because properties of the nanocrystalline-TiO_2 layer [pore size, diffusion of materials (electrolytes), surface area, dye adsorption, light diffraction (haze), and light penetration] can change with the change in TiO_2 nanoparticle size, as shown in Figure 4.[15] For optimization using nanocrystalline-TiO_2 solar cells, each parameter and physical meaning has been changed, which is thought to consider every parts. Hence, in some cases, just attempting to fabricate nanocrystalline-TiO_2 solar cells and to check the photovoltaic results can be the straightforward way to understand the phenomena.

When the polymer/TiO_2 ratio in the TiO_2-coating paste remained unchanged, the pore size increased linearly with the increase in particle size (Figure 5). However, if the polymer/TiO_2 ratio in the TiO_2-coating paste was changed, the pore size can change without changing the particle size (Figure 6). On changing binder polymer amounts from 30 to 50%, the peak was shifted to larger pore size. However, no change in the peak position was observed when binder polymer amounts were changed from

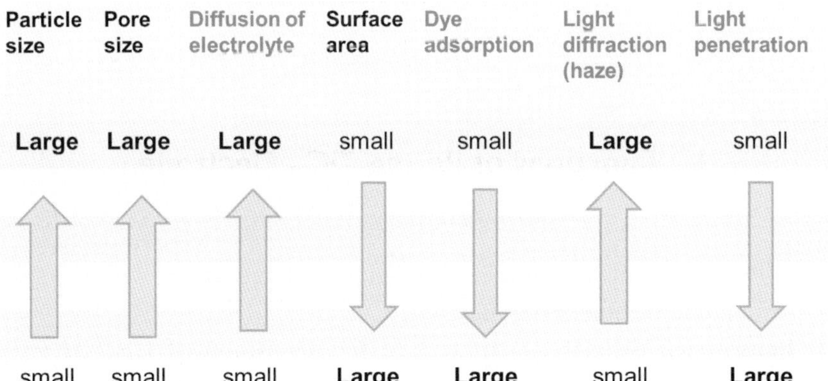

Figure 4. Influences of nanocrystalline-TiO_2 electrodes (particle size). Adapted from Ref. 15.

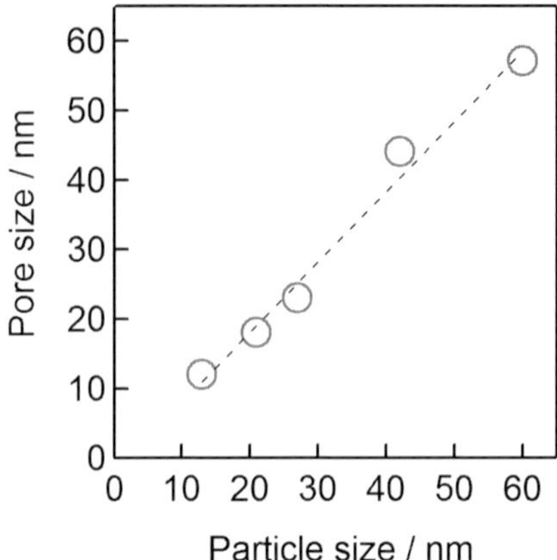

Figure 5. Relationship between particle and pore sizes with keeping the ratio of TiO$_2$ and polymer amounts. Adapted from Ref. 15.

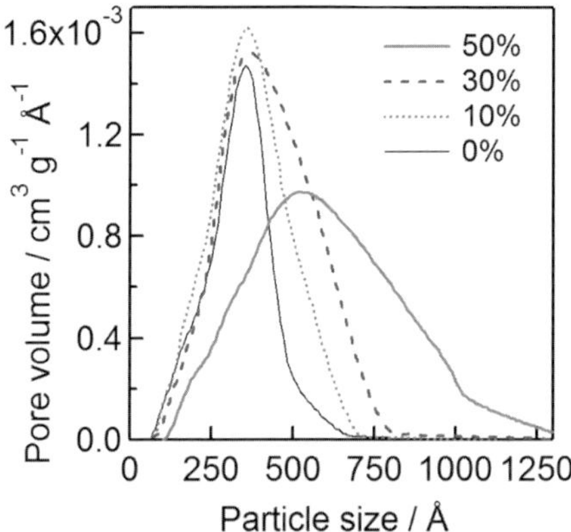

Figure 6. Pore-size distributions with different amounts of polymer binder in TiO$_2$ pastes. The numbers in figure show the amount of polymer binder. Adapted from Ref. 16.

0 to 30%, but the distribution become wider to larger pores with increasing the amount of binder polymer.[16]

In two-step (sequential) procedure for deposition of CH$_3$NH$_3$PbI$_3$ perovskite crystals,[17] porous-TiO$_2$ electrodes using smaller nanoparticle kept much amount of PbI$_2$, which was confirmed by X-ray diffraction

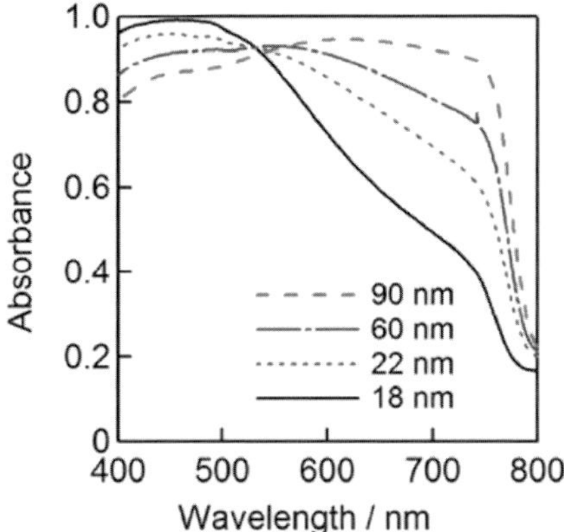

Figure 7. Reflectance absorption spectra of <glass/nanocrystalline-TiO$_2$/CH$_3$ NH$_3$PbI$_3$ perovskite>. The numbers in the figure show the diameters of TiO$_2$ nanoparticles.

(XRD).[18] This phenomenon can also be confirmed by reflectance absorption spectra (Figure 7). The TiO$_2$ pore sizes for the nanocrystalline-TiO$_2$ layers were related to the data shown in Figure 5. Keeping the isosbestic point at 520 nm, the smaller TiO$_2$ nanoparticles increased the peak at 460 nm for PbI$_2$ and decreased the absorption from 550 to 800 nm for CH$_3$NH$_3$PbI$_3$ perovskite crystal.

Table 1 shows the best photovoltaic parameters of CH$_3$NH$_3$PbI$_3$ perovskite solar cells with different cell fabrication methods, different structures, and different TiO$_2$ nanoparticle sizes. The pore sizes for the nanocrystalline-TiO$_2$ layers were related to the data shown in Figure 5. Data were the average of three cells. Because data were picked up from the best value varied by the particle size, the photovoltaic parameters [short-circuit photocurrent density (J_{sc}), open-circuit photovoltage (V_{oc}), fill factor (FF), and PCE] were not related to each other. The structures of two perovskite solar cells are shown in Figure 8. Figure 8a shows the normal structure of organolead halide perovskite solar cells, which contains vacuum-evaporated gold layer for the back contact and nanocrystalline-TiO$_2$ layer on dense (or, compact, blocking) TiO$_2$ layer. The dense TiO$_2$ layer is important to block short circuit between perovskite and F-doped tin oxide (FTO).

Table 1. Best photovoltaic parameters of $CH_3NH_3PbI_3$ perovskite solar cells with different schemes, different structures, and different TiO_2 nanoparticle sizes. Data were the average of three cells. The numbers in parentheses show the particle diameter at the best parameter data. Because data were picked up from the best value varied by the particle size, the photovoltaic parameters (J_{sc}, V_{oc}, FF, and PCE) were not related to each other. Adapted from Refs. 18 and 21.

Cell information	J_{sc}/mA·cm^{-2}	V_{oc}/V	FF	PCE/%
<porous-TiO$_2$/ CH$_3$NH$_3$PbI$_3$/ CuSCN/Au> by one-step method	19.88 (60 nm)	0.75 (36 nm)	0.60 (60 nm)	6.34 (36 nm)
<porous-TiO$_2$/ CH$_3$NH$_3$PbI$_3$/ CuSCN/Au> by two-step method	19.84 (36 nm)	0.70 (60 nm)	0.62 (18 nm)	7.68 (36 nm)
<porous-TiO$_2$ + CH$_3$NH$_3$PbI$_3$/ porous-ZrO$_2$ + CH$_3$NH$_3$PbI$_3$/ porous-carbon + CH$_3$NH$_3$PbI$_3$> by two-step method	7.27 (45 nm)	0.90 (81 nm)	0.62 (81 nm)	3.02 (13 nm)

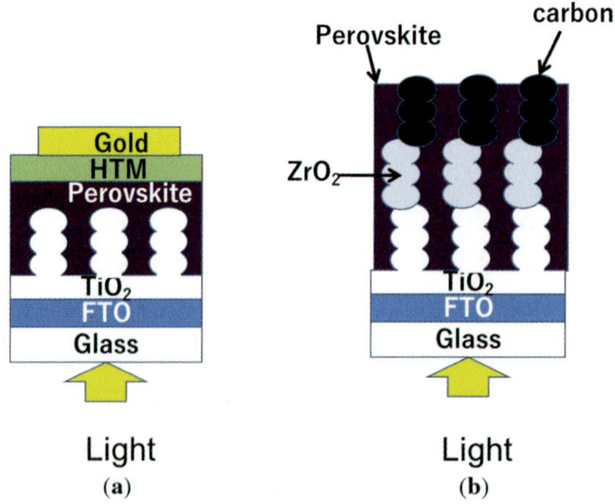

Figure 8. Structures of perovskite solar cells: (a) normal organolead halide perovskite solar cells and (b) fully printed HTL-free perovskite solar cells. "HTM" and "FTO" are a hole-transporting material and fluorine-doped tin oxide, respectively.

Figure 8b shows the fully printed hole-transporting material (HTM)-free perovskite solar cells investigated by Professor Hongwei Han (Huazhong University of Science and Technology, China) and coworkers.[19] Although the stability of normal structure perovskite solar cells (Figure 8a) has been an important issue to be solved, the stability of fully printed HTL-free perovskite solar cells (Figure 8b) is quite promising, which can survive under 1-sun illumination (AM 1.5G, 100 mW·cm^{-2}) without encapsulation over 1000 h[19] and at 100 °C in dark over 1000 h with encapsulation.[21]

It can be noticed that the best photovoltaic parameters were obtained with different TiO_2 sizes. Although the PCE for <porous-TiO_2/$CH_3NH_3PbI_3$/CuSCN/Au> structure solar cells (Figure 8a) had the peak at 36 nm TiO_2 diameter by one-step and two-step $CH_3NH_3PbI_3$ deposition, the TiO_2 diameter varied for other parameters. Moreover, the smaller TiO_2 nanoparticle (as 13 nm) looks better for the solar-cell structure of <porous-TiO_2 + $CH_3NH_3PbI_3$/porous-ZrO_2 + $CH_3NH_3PbI_3$/porous-carbon + $CH_3NH_3PbI_3$> (Figure 8b). The photovoltaic parameters were affected by the parameters of nanocrystalline-TiO_2 layers listed in Figure 4. Each parameter was connected, intricately. In this case, specially, material diffusion to the bottom of porous layer may be a quite important point. The conversion efficiency data of fully printed HTM-free perovskite solar cells <porous-TiO_2 + $CH_3NH_3PbI_3$/porous-ZrO_2 + $CH_3NH_3PbI_3$/porous-carbon + $CH_3NII_3PbI_3$> (in Table 1) were quite lower than those in the publication from Professor Hongwei Han and coworkers (12.8% in Ref. 19). In this case, therefore, porous-TiO_2, porous-ZrO_2, and porous-carbon layers should be optimized, and the perovskite deposition method also should be improved.

IV. Electron Transportation and Accumulation on the TiO_2 Surface

The second important point to understand the function of nanocrystalline-TiO_2 electrode in dye-sensitized and perovskite solar cells is the ability of TiO_2 to transport and accumulate electrons on TiO_2 surface (as Figure 9).

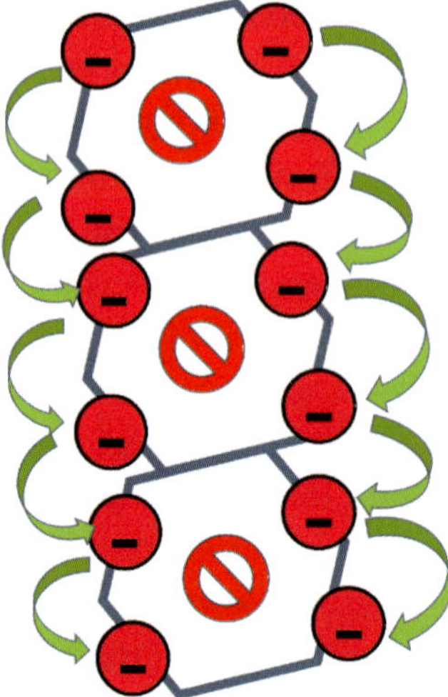

Figure 9. Schematic image of electron transfer *via* surfaces of TiO$_2$ nanoparticles.

Such electron accumulation on TiO$_2$ surface was confirmed by Zhu *et al.*[22] They have confirmed this phenomenon by measurements of TiO$_2$ diffusion coefficient and electron density on TiO$_2$ surface with changing the TiO$_2$ nanoparticle sizes. This accumulation and transportation of electrons on surface can be related to the low-temperature-annealed porous nanocrystalline-TiO$_2$ electrode for dye-sensitized solar cells, because electrons can hop to the next TiO$_2$ nanoparticles *via* connected surfaces of particles. If the electron is inside the crystal, the nonheated surface can be just a barrier for hopping. For this electron hop *via* TiO$_2$ surface, the charge shielding by adsorbed cation on the TiO$_2$ surface can support the transfer. Without cation in the electrolyte, electrons on TiO$_2$ surface in dye-sensitized solar cells could not move (Figure 10). However, the increase of cation increased the photoresponse of nanocrystalline-TiO$_2$ electrode.[23] This charge shielding effect is also a kind of evidence of electron accumulation on the surface of TiO$_2$ nanoparticles. It can be considered that the electron transfers on TiO$_2$ surface in organolead halide perovskite solar cells have been assisted by ion migration in the perovskite crystal.[24,25]

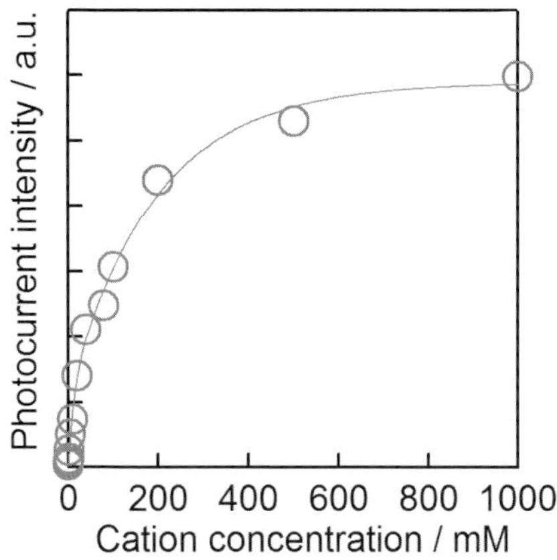

Figure 10. Photocurrent response at laser pulse irradiation (Quanta-Ray, INDI Series Pulsed Nd:YAG lasers; pulse width: 7 ns; wavelength: 355 nm, with average electron density of $2.1 \times 10^{17} \cdot cm^{-3}$) on the surface of 7.2-$\mu$m-thick nanocrystalline-$TiO_2$ layer with various concentrations of $TBA^+ClO_4^-$ electrolyte in acetonitrile. Adapted from Ref. 23.

V. Prohibition of Short Circuit between Working and Counter Electrodes

The third point of nanocrystalline-TiO_2 function is prohibition of short circuit between working electrode and counter electrodes. A typical and practical structure of dye-sensitized solar cells is shown in Figure 11. For the experiments in laboratory, a dye-sensitized solar cell can be fabricated by assembling a working electrode (dye-adsorbed porous-TiO_2 layer on FTO-coated glass substrate) and a counter electrode (Pt-loaded FTO-coated glass substrate) using clips, and then, iodine electrolyte can be introduced between the electrodes for the activation of dye-sensitized solar cell. The surfaces of porous-TiO_2 working electrodes and Pt counter electrodes attach directly to each other. However, physical scientists have been surprised at this moment, because such direct contact between working and counter electrodes must cause short circuit. With this strange structure, dye-sensitized solar cells can work properly (as shown in Figure 12).[26] It was confirmed that a partial short circuit occurred in dye-sensitized solar cells, but the most of exited careers in the nanocrystalline-TiO_2 electrode can be transferred to the FTO/glass substrate of working electrodes

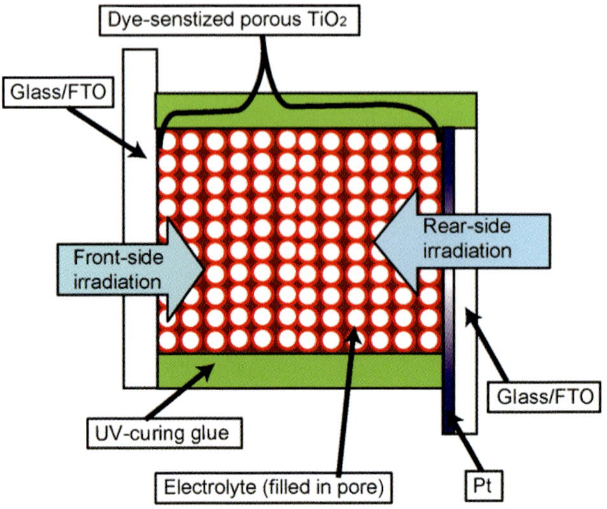

Figure 11. Schematic cell structure of bifacial dye-sensitized solar cells. Adapted from Ref. 26.

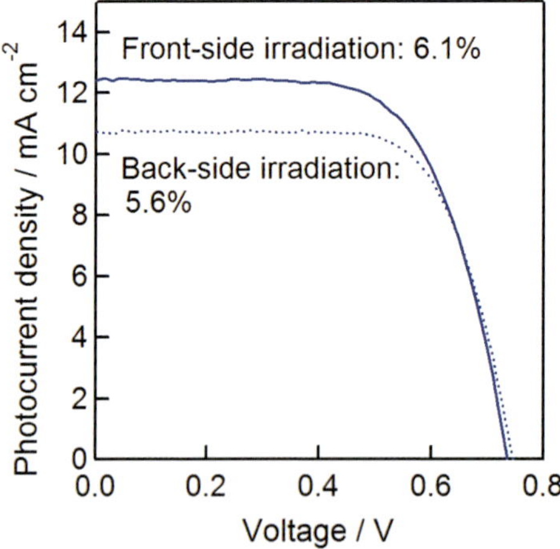

Figure 12. Photocurrent–voltage curves of bifacial dye-sensitized solar cells. Adapted from Ref. 26.

without charge recombination.[26] This is a very interesting phenomenon of nanocrystalline-TiO_2 electrode in dye-sensitized solar cells.

For the organolead halide perovskite solar cells, this phenomenon of nanocrystalline-TiO_2 electrode can help to prevent the short circuit

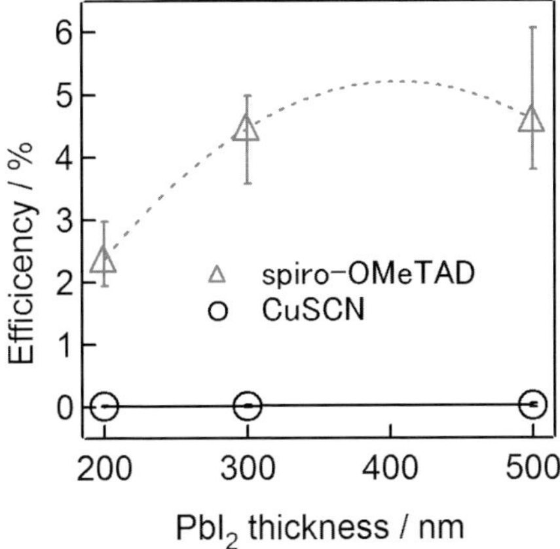

Figure 13. Variation of PCE of <FTO/flat-TiO₂/CH₃NH₃PbI₃/CuSCN/Au> and <FTO/flat-TiO₂/CH₃NH₃PbI₃/spiro-OMeTAD/Au> cells by two-step-processed CH₃NH₃PbI₃ perovskite related to the thickness of the PbI₂ layer on flat-TiO₂ electrodes. Adapted from Ref. 18.

between working and counter electrodes. To fabricate planar perovskite solar cells by two-step (sequential) deposition method of $CH_3NH_3PbI_3$ perovskite layer,[17] two kinds of hole-transporting layers have been tried: one is spiro-OMeTAD (2,2′,7,7′-tetrakis-(N,N-di-p-methoxyphenylamine)-9,9′-bifluorene; a standard organic hole-transporting material for perovskite solar cells)[17] and the other is CuSCN (copper thiocyanate; a cost-effective inorganic hole-transporting material).[27] Using porous nanocrystalline-TiO₂ electrodes, CuSCN had worked properly as perovskite solar cells (as shown in Table 1). Although spiro-OMeTAD can work properly to be planar perovskite solar cells, however, CuSCN could not give photovoltaic effect at all (Figure 13).[18] To prohibit the short circuit by thicker $CH_3NH_3PbI_3$ (or PbI_2) layer in the two-step method, the thickness of PbI_2 was varied by spin-coating acceleration, but such thicker PbI_2 (and the resulting $CH_3NH_3PbI_3$ perovskite) layer cannot improve the photovoltaic effect. It was considered that the short circuit is due to the interdiffusion between $CH_3NH_3PbI_3$ perovskite and CuSCN layers[18,27,28] (Figure 14). The interdiffusion can attach the CuSCN HTM directly on the dense TiO₂ layer in the planar perovskite solar cells (Figure 14a), which can cause short circuit. On the contrary, although CuSCN can attach

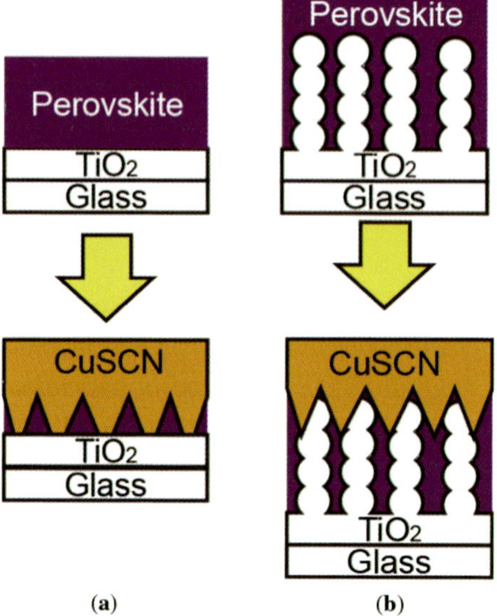

Figure 14. Schematic images of CuSCN diffusion into organometal halogen perovskite layer: (**a**) on planar-TiO_2 electrode and (**b**) on porous-TiO_2 electrode.

porous-TiO_2 layer in the mesoscopic-structure perovskite solar cells (Figure 14b), this porous nanocrystalline-TiO_2 electrode can prohibit short circuit and keep the photovoltaic activity for the perovskite solar cells using CuSCN HTM (as shown in Table 1).

VI. Conclusions and Perspectives

The history of dye sensitization (to improve the photoactivation of nanocrystalline TiO_2) and the function of nanocrystalline-TiO_2 electrodes have been reviewed. Specially, the three important functions are as follows: (1) tuning of particle and hole size for light transparency and material diffusion, (2) electron transportation and accumulation on the surface of TiO_2 nanoparticles, and (3) prohibition of short circuit. Although planar perovskite solar cells have been fabricated, the best efficiency perovskite solar cells are still fabricated using porous nanocrystalline-TiO_2 electrodes. Therefore, knowledge of the three functions of nanocrystalline-TiO_2 electrodes is important for further improvement in this field.

VII. References

1. Fujishima, A.; Honda, K. *Nature* **1972**, *238*, 37–38.

2. Gratzel, M. *Cattech* **1999**, *3*, 4–17.

3. Bibliotheca Novissima observationum et recensionum. Ed. J. Chr. Franck, Sectio V, Nr. VII. Halae Magdeburgicae 1719, S. 234–240; Acta physico-medica, Leopoldina Band 1, 1727, S. 528–532; Leslie Stroebel and Richard D. Zakia (1993). The Focal encyclopedia of photography (3rd ed.). Focal Press. p. 6. ISBN 978-0-240-51417-8; Susan Watt (2003). Silver. Marshall Cavendish. pp. 21–. ISBN 978-0-7614-1464-3; Litchfield, Richard Buckley (1903). Tom Wedgwood, the First Photographer, etc., London, Duckworth and Co. Out of copyright and available free at archive.org. In Appendix A (pp. 217–227); Niépce House Museum: Invention of Photography: 1816-1818, Niépce's first tries; < https://ja.wikipedia.org/wiki/ ; https://en.wikipedia.org/wiki/Johann_Heinrich_Schulze >

4. Jessica Gorman (2007). "Photography at a Crossroads". Science News 162 (21): 331–333. doi:10.2307/4013861. JSTOR 4013861; Around the World in 1896: A Brief History of Photography." The Library of Congresss. 2002. 18 Sep 2008; Baatz, Willfried (1997). Photography: An Illustrated Historical Overview. New York: Barron's. p. 16. ISBN 0-7641-0243-5; "The 150th Anniversary of Photography," in History of Photography, Vol. I, No. 1, January 1977; Marilyn Stokstad, David Cateforis, Stephen Addiss (2005). Art History (Second ed.). Upper Saddle River, New Jersey: Pearson Education. pp. 964. ISBN 0-13-145527-3 < https://ja.wikipedia.org/wiki/ ; https://en.wikipedia.org/wiki/ Nicéphore_Niépce >

5. Phil Coomes (27 April 2010). "Remembering Frederick Scott Archer". BBC. Retrieved 2010-12-06; "Frederick Scott Archer". British Journal of Photography. 22 (773): 102–104. 26 February 1875; Peres, Michael R., ed. (2007). Focal Encyclopedia of Photography: Digital Imaging, Theory and Applications. Focal Press. p. 124. ISBN 978-0-240-80740-9. < https://en.wikipedia.org/wiki/Frederick_Scott_Archer >

6. Vogel, H., "On the sensitiveness of bromide of silver to the so-called chemically inactive colours", Chemical News, December 26, 1873:318–319, copying from The Photographic News, date and page not cited but apparently December 12, 1873; Vogel, H., "Photo-spectroscopic researches", The Photographic News, March 20, 1874:136–137; Vogel, H., "Rendering actinic non-actinic rays", The Photographic News, July 3, 1874:320–321 < https://en.wikipedia.org/wiki/Hermann_Wilhelm_Vogel >

7. Moser, J. *Monatsh. Chem.* **1887**, *8*, 373.

8. Tsubomura, H.; Matsumura, M.; Nomura, Y.; Amamiya, T. *Nature* **1976**, *261*, 402–403.

9. O'Regan, B.; Grätzel, M. *Nature* **1991**, *353*, 737–740.

10. Grätzel, M. *Chem. Lett.* **2005**, *34*, 8–13.

11. Lee, M. M.; Teuscher, J.; Miyasaka, T.; Murakami, T. N.; Snaith, H. J. *Science* **2012**, *338*, 643–647.

12. Kim, H.-S.; Lee, C.-R.; Im, J.-H.; Lee, K.-B.; Moehl, T.; Marchioro, A.; Moon, S.-J.; Humphry-Baker, R.; Yum, J.-H.; Moser, J. E.; Grätzel, M.; Park, N.-G. *Sci. Rep.* **2012**, *2*, 591.

13. Jeon, N. J.; Noh, J. H.; Yang, W. S.; Kim, Y. C.; Ryu, S.; Seo, J.; Seok, S. I., *Nature* **2015**, *517*, 476–480.

14. Yang, W. S.; Noh, J. H.; Jeon, N. J.; Kim, Y. C.; Ryu, S.; Seo, J.; Seok, S. I. *Science* **2015**, *348*, 1234.

15. Ito, S.; Murakami, T. N.; Zakeeruddin, S. M.; Yazawa, T.; Mizuno, M.; Kayama, S.; Grätzel, M. *Nano* **2014**, *09*, 1440010.

16. Barbé, C. J.; Arendse, F.; Comte, P.; Jirousek, M.; Lenzmann, F.; Shklover, V.; Grätzel, M., *J. Am. Ceram. Soc.* **1997**, *80*, 3157–3171.

17. Burschka, J.; Pellet, N.; Moon, S. J.; Humphry-Baker, R.; Gao, P.; Nazeeruddin, M. K.; Graetzel, M. *Nature* **2013**, *499*, 316–319.

18. Murugadoss, G.; Mizuta, G.; Tanaka, S.; Nishino, H.; Umeyama, T.; Imahori, H.; Ito, S., APL *Materials* **2014**, *2*, 081511.

19. Mei, A.; Li, X.; Liu, L.; Ku, Z.; Liu, T.; Rong, Y.; Xu, M.; Hu, M.; Chen, J.; Yang, Y.; Grätzel, M.; Han, H. *Science* **2014**, *345*, 295.

20. Ito, S.; Mizuta, G.; Kanaya, S.; Kanda, H.; Nishina, T.; Nakashima, S.; Fujisawa, H.; Shimizu, M.; Haruyama, Y.; Nishino, H.; *Phys.Chem.Chem.Phys.*, **2016**, *18*, 27102–27108.

21. Baranwal, A. K.; Kanaya, S.; Peiris, T. A. N.; Mizuta, G.; Nishina, T.; Kanda, H.; Miyasaka, T.; Segawa, H.; Ito, S. *ChemSusChem*, **2016**, *9*, 2604–2608.

22. Zhu, K.; Kopidakis, N.; Neale, N. R.; van de Lagemaat, J.; Frank, A. J. *J. Phys. Chem. B* **2006**, *110*, 25174–25180.

23. Kambe, S.; Nakade, S.; Kitamura, T.; Wada, Y.; Yanagida, S. *J. Phys. Chem. B* **2002**, *106*, 2967–2972.

24. Yuan, Y.; Chae, J.; Shao, Y.; Wang, Q.; Xiao, Z.; Centrone, A.; Huang, J. *Adv. Energy Mater.* **2015**, *5*, 1500615.

25. Eames, C.; Frost, J. M.; Barnes, P. R. F.; O'Regan, B. C.; Walsh, A.; Islam, M. S. *Nat. Commun.* **2015**, *6*, 7497.

26. Ito, S.; Zakeeruddin, S. M.; Comte, P.; Liska, P.; Kuang, D.; Gratzel, M. *Nat. Photon.* **2008**, *2*, 693–698.

27. Qin, P.; Tanaka, S.; Ito, S.; Tetreault, N.; Manabe, K.; Nishino, H.; Nazeeruddin, M. K.; Grätzel, M. *Nat. Commun.* **2014**, *5*, 3834.

28. Ito, S.; Kanaya, S.; Nishino, H.; Umeyama, T.; Imahori, H. *Photonics* **2015**, *2*, 1043.

4 P-Type and Inorganic Hole Transporting Materials for Perovskite Solar Cells

Ming-Hsien Li, Yu-Hsien Chiang, Po-Shen Shen, Sean Sung-Yen Juang and Peter Chao-Yu Chen*

Department of Photonics, National Cheng Kung University, Tainan 701, Taiwan
*petercyc@mail.ncku.edu.tw

Table of Contents

List of Abbreviations

BCP	bathocuproine
CB	conduction band
CuI	copper iodide
CuInS$_2$	copper indium disulfide
Cu$_2$O	cuprous oxide
CuO	cupric oxide
CuSCN	copper thiocyanate
CuS	copper sulfide
CIGS	copper indium gallium selenide
DSC	dye-sensitized solar cell
ETM	electron transporting material
FDC	fast deposition-crystallization
FF	fill factor
FTO	fluorine-doped tin oxide
HTM	hole transporting material
ITO	indium-doped tin oxide
J_{SC}	short circuit current density
LUMO	lowest unoccupied molecular orbital
MAI	ammonium iodide
NiO	nickel oxide
OPV	organic photovoltaic
PHJ	planar heterojunction junction
PCE	power conversion efficiency
PIA	photoinduced transient absorption
PSC	perovskite solar cell

Psk	perovskite
PCBM	phenyl-C_{61}-butyric acid methyl ester
$PC_{60}BM$	phenyl-C_{60}-butyric acid methyl ester
PEDOT	poly(3,4-ethylenedioxythiophene)
PSS	poly(styrene sulfonate)
Spiro-OMeTAD	2,2′,7,7′-tetrakis-(*N*,*N*-di-*p*-methoxyphenylamine)-9,9′-bifluorene
TCO	transparent conductive oxide
TRTS	time-resolved terahertz spectroscopy
VB	valence band
V_{OC}	open circuit voltage
XPS	X-ray photoelectron spectra

I. Introduction

The emerging solid-state organometallic lead halide perovskite solar cells have attracted tremendous attention because of their promising power conversion efficiencies (PCEs) which have been boosted to over 22% in 2016.[1] Basically, hybrid organic–inorganic perovskite with a crystalline structure of ABX_3, where A is an organic cation [e.g., methylammonium ($CH_3NH_3^+$) or formamidinium ($CH(NH_2)_2^+$)], B is a metal cation (typically Sn^{2+} or Pb^{2+}), and X is a halide anion (such as Cl^-, Br^-, and I^-), consists of a three-dimensional (3D) framework of corner-sharing BX_6 octahedron with the A ion placed in the cuboctahedral interstices. The most commonly used hybrid organic–inorganic perovskite $CH_3NH_3PbI_3$ possess promising properties for solar cells including excellent voltage output (~1.0 V) w.r.t to band gap (1.5 eV), high extinction coefficient (~10^4 cm^{-1} at 550 nm) and broad absorption range from visible (vis) to near-infrared spectrum (300–800 nm),[2] low exciton binding energy (BE) (20–30 meV),[3] long and relatively balanced electron and hole diffusion lengths,[4] ambipolar carrier transport capabilities with high electron and hole mobility of 5–12 and 1–8 cm^2/Vs, respectively,[5] and pronounced defect tolerance. $CH_3NH_3PbI_3$ and $CH_3NH_3PbI_{3-x}Br_x$ were first employed to fabricate sensitized-type solar cells with iodide liquid electrolyte by Miyasaka's group,[6] and a PCE beyond 6% was further improved by modifying the TiO_2 surface and the processing method for the perovskite deposition.[7]

However, the perovskite materials are not stable in highly polar liquid electrolyte. A remarkable success for perovskite solar cell (PSC) was obtained by replacing the liquid electrolyte with solid-state hole conductors. The first realized solid-state perovskite-based solar cells using spiro-OMeTAD (2,2′,7,7′-tetrakis-(*N,N*-di-*p*-methoxyphenylamine)-9,9′-bifluorene) as hole transporting materials (HTMs) were fabricated with mesoscopic TiO_2 and super-mesostructure Al_2O_3 scaffold, achieving a PCE exceeding 9%[8] and over 10%,[9] respectively. As evolved from solid-state dye-sensitized solar cells (DSCs), the typical PSC displays a device architecture that is composed of a compact hole-blocking layer deposited on a transparent conductive oxide (TCO) substrate to suppress the recombination between carriers from the perovskite and TCO, followed by a mesoporous metal oxide as the electron-transporting material (ETM) or the insulating scaffold to provide porous matrix to adsorb perovskite. The perovskite is then filled on the mesoporous layer with a capping layer, followed by the deposition of HTM to extract and transfer holes from the photoexcited perovskite. Finally, thermally deposited silver or gold is covered on HTM to complete the device. Figure 1a schematically presents the regular structure of DSC-like PSC. With the evolution from mesoscopic sensitized junction to noninjecting super mesostructure scaffolds, planar heterojunction (as referred to Figure 1b) by elimination of mesoscopic scaffold was further proposed and became a popular device configuration, which simplifies device fabrication process with reduced material cost and sintering temperature. An alternative configuration with the illumination from the p-type substrate was developed as inverted structure. The PSC based on p-i-n heterojunction is revealed as organic photovoltaic (OPV)-like structures, as shown in Figure 1c and 1d. The inverted p-i-n configuration facilitates charge-generation profile with more effective hole extraction to the TCO and suppress hysteresis behavior compared to the regular n-i-p heterojunction. Furthermore, it provides us more choices on the designs for integration of PSCs with silicon or copper indium gallium selenide (CIGS) solar cells to build tandem devices.[10]

Since the first successful application of spiro-OMeTAD in solid-state PSCs, it has been extensively employed for fabrication of high-efficiency PSCs.[11] Although spiro-OMeTAD has been demonstrated as an excellent small molecule HTM, its high production cost accompanied

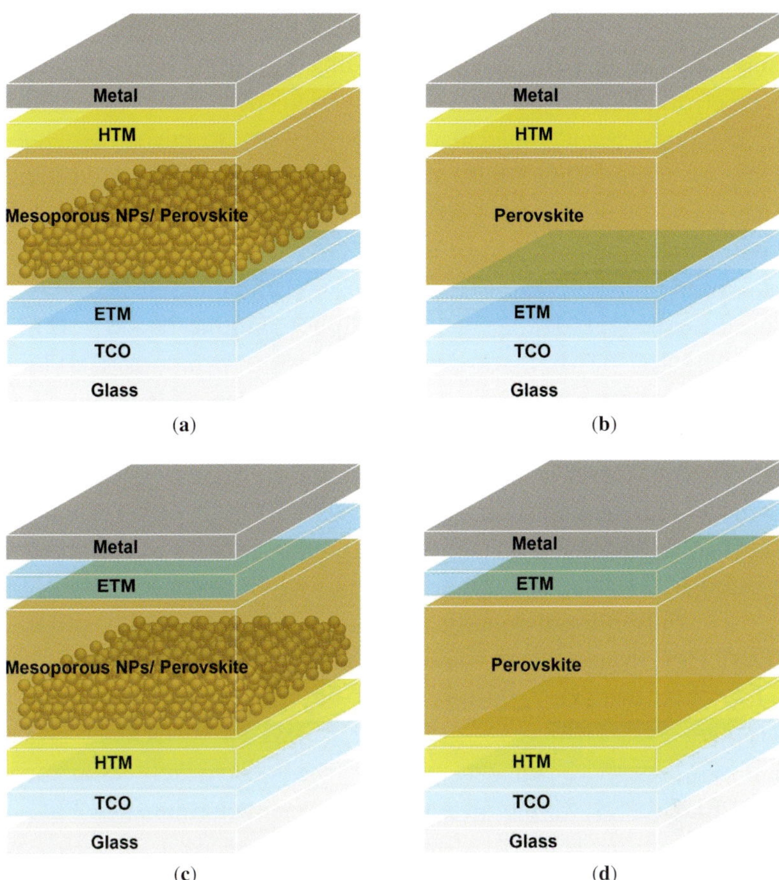

Figure 1.   Schematic illustration of PSCs devices: (**a**) n-i-p mesoscopic architecture; (**b**) n-i-p planar architecture; (**c**) p-i-n mesoscopic architecture; and (**d**) p-i-n planar architecture. Taken from Ref. 10b and Ref. 13 with permission of MDPI AG and the Royal Society of Chemistry.

by multistep synthesis and tedious purification, lower hole mobility, and lack of long-term stability curtail the further viability of PSCs. For the commercial purposes, several efficient HTMs have been proposed to replace spiro-OMeTAD aiming for low cost and stability. Owing to the wide varieties of HTMs, the p-type contact materials for PSCs can be categorized into organic,[10d,12] inorganic,[10b,13] and organometallic HTMs.[14] Organic HTMs, including small molecular and polymeric

HTMs, are low-cost alternatives capable of low-temperature device processing. They are advantageous in providing diversified physical and optical properties to match the band gaps of various perovskites by modifying their structures. Small molecular HTMs are easy to purify, suitable for forming crystalline thin films, and feasible to match the band gaps of various perovskites by modifying their structures. Various small-molecule p-type contact, including spiro-based HTMs,[15] spiro-OMeTAD derivatives,[16] bimesitylene-based HTM,[17] triphenylamine-based HTMs,[18] carbazole-based HTMs,[19] methoxydiphenylamine-based HTM,[20] thiophene-based HTMs,[21] star-shaped HTMs,[21a,22] pyrene-core arylamine derivatives,[23] tetrathiafulvalene derivatives,[24] perylene,[25] BuPyIm-TFSI,[26] triarylamine-based HTM,[27] truxene derivatives,[28] phenoxazine-based HTM,[29] and asymmetric squaraine-based HTM,[30] have been developed to show decent efficiencies between 11 and 20%. On the other hand, polymeric HTMs present good stability and processibility, and are applied to fabricate PSCs with decent efficiency. PSCs made with polymeric HTMs, such as poly(3,4-ethylenedioxythiophene):poly(styrene sulfonate) (PEDOT:PSS) based HTM,[31] oligothiophene derivatives,[32] DPP-based polymer,[33] polyTPD,[34] VB-DAAF,[35] PDPPDBTE,[36] PCPDTBT,[37] PCDTBT,[37a,38] PTAA,[37a,38,39] PDPP3T,[40] P3TAA,[41] DOR3T-TBDT,[42] and P3HT,[43] have shown decent PCE. The most widely used organic HTM, PEDOT:PSS, for the inverted p-i-n planar heterojunction PSCs has been reported to achieve a recorded PCE of 18.1%.[44] However, PEDOT:PSS-based devices display inferior reliability and stability because of its acidity, high water absorption, and incapacity to block electrons. The use of organic hole conductors may cause a potential hurdle to the future commercialization of PSC from the consideration of cost and stability. Alternatively, inorganic p-type semiconductors have attracted intense attention as ideal candidates due to their low cost, ease of synthesis, high hole mobility benefiting hole collection, high transmittance in the vis region, and good chemical stability. These wide band gap HTMs possess deep valance band (VB) position to produce high open circuit voltage (V_{OC}) and high conduction band (CB) edges to effectively block electron transporting to anode. Accordingly, PSCs employing inorganic HTMs have been demonstrated to exhibit decent stability. Furthermore, the use of inorganic charge transport layer of ETM and HTM for PSCs has been demonstrated to protect the perovskite photoactive layer from exposure to ambient environments, thus enhancing the resistance to degradation of perovskite and achieving highly stable perovskite-based solar cells.[45] Briefly, ideal HTMs for highly efficient PSCs are required to offer the following features: appropriate CB for efficient electron blocking and VB for hole collection, high hole mobility, stable thermal and optical properties. The scope of this chapter is to summarize the developments in PSCs using inorganic HTMs as the p-type contact. Based on the contact electrode polarity, DSC-like structures that work with n-i-p heterojunction are firstly introduced, followed by the OPV-like inverted structures of p-i-n heterojunction, in which the positions of ETM and HTM are inverted. For both DSC-like and OPV-like architecture, mesoscopic and planar PSCs are also reviewed.

II. Inorganic HTM of NiO for PSCs

Nickel oxide (NiO) is a low-cost material with superior properties, such as thermal and chemical stability, high optical transparency with a wide band gap of 3.5–3.9 eV, which has been demonstrated to be a good candidate of hole selective contact for PSCs in virtue of its suitable work function of about −5.3 eV and suitable CB position to effectively transport holes and block electrons. A significant quenching of photoluminescence emission from NiO/perovskite heterojunction confirms that NiO film effectively extracts holes from perovskite, though the first planar architecture of $NiO/CH_3NH_3PbI_{3-x}Cl_x/PCBM$ obtained low efficiency (<0.1%) mainly due to very poor perovskite coverage on the NiO surface.[46] Since then, many efforts have been devoted to improve the device efficiency using NiO as hole transporter. The details are discussed by various device configurations.

A. Mesoporous NiO-Based n-i-p PSCs

Liu *et al.* introduced NiO nanoparticles to form a mesoporous NiO space separator between the mesoporous TiO_2 and the carbon counter electrode shown in Figure 2a. A hetero p-n Schottky barrier formed by NiO/TiO_2 separates electron and hole to flow in opposite direction. The PSCs were

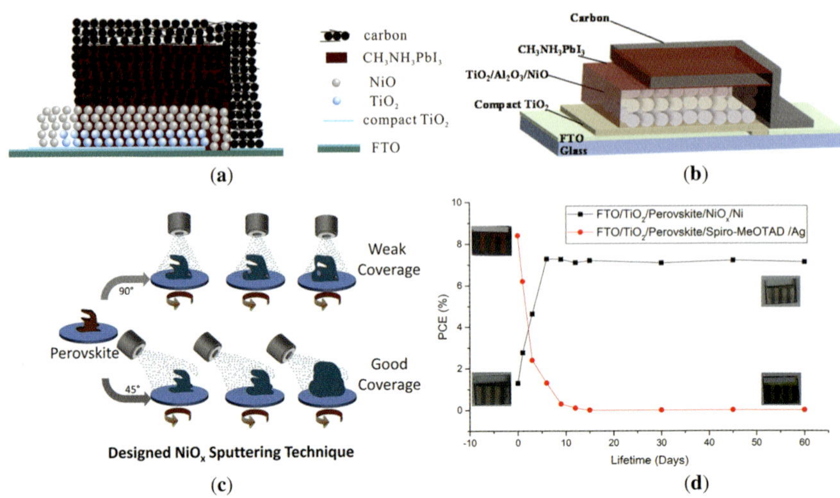

Figure 2. (**a**) Schematic representation of the mp-TiO_2/mp-NiO/Carbon/perovskite structure. Taken from Ref. 48 with permission of the Royal Society of Chemistry. (**b**) Schematic representation of the mp-TiO_2/mp-Al_2O_3/mp-NiO/Carbon/perovskite structure. Taken from Ref. 51 with permission of Elsevier Science. (**c**) Schematic representation of the 45° deposition of NiO on the perovskite. (**d**) Device performance durability of the TiO_2/perovskite/NiO cell for a period of 60 days. Taken from Ref. 45d with permission of the American Chemical Society.

composed of fluorine-doped tin oxide (FTO)/compact TiO_2/mesoporous TiO_2/mesoporous NiO/carbon, in which the light absorber $CH_3NH_3PbI_3$ was infiltrated inside the pores of mesoporous layers *via* sequential deposition.[47] Application of mesoporous NiO as HTMs effectively retards charge recombination and facilitates hole extraction, resulting in a cell with PCE of 11.4%.[48] Furthermore, the same groups applied highly crystalline NiO nanosheets as top nanostructured charge transport layers and inserted an additional mesoporous ZrO_2 layer to achieve a n-i-p heterojunction. The ZrO_2 layer was inserted to separate the mesoporous TiO_2 layer and the mesoporous NiO layer to further suppress TiO_2/NiO interface charge recombination, leading to an enhancement of charge collection. A full device was fabricated as a structure of FTO/compact TiO_2/mesoporous TiO_2/mesoporous ZrO_2/mesoporous NiO/carbon/sequential deposited $CH_3NH_3PbI_3$. The PSCs using all metal oxide semiconductor as a framework achieved a promising energy conversion efficiency of 14.2%.[49] Similar works were reported by Xu *et al.* who used a doctor blade technique to fabricate mesoscopic TiO_2/ZrO_2/NiO/carbon structures that exhibited an appreciated PCE of 14.9%.[50] Under a similar configuration, a quadruple layer consisted of mesoscopic TiO_2/Al_2O_3/NiO/carbon was employed as a scaffold and deposited by screen printing to fabricate mesoscopic PSCs (referred to Figure 2b). By optimizing the thickness of each layer, a considerable PCE of 15.03% was achieved for 800-nm-thick mesoporous NiO layer.[51] The abovementioned mesoscopic devices displayed excellent stability due to the protection provided by the thick carbon counter electrode.

B. Planar NiO-Based n-i-p PSCs

Nejand *et al.* used a sputter-deposited NiO_x thin film on the perovskite light absorber to extract the photo-generated holes from the perovskite layer. A nickel electrode was subsequently sputtered to replace the noble Au contact due to its vicinity of work function close to the VB of perovskite. A planar n-i-p heterojunction PSC of FTO/compact TiO_2/ $CH_3NH_3PbI_{3-x}Cl_x$/NiO_x/Ni was fabricated. The substrate was tilted 45° against the sputtering target for the formation of compact and uniform NiO_x layer covering on perovskite surface as seen in Figure 2c. Also, periodic deposition with controlled temperature was further conducted to lower the substrate temperature without damaging the perovskite layer. The fabricated device initially displayed a low photovoltaic performance. After the strain release of the sputter-deposited NiO_x layer, high hole mobility in NiO_x film was achieved and led to improvement in both V_{OC} and J_{SC}. Eventually, PSCs employing sputter-deposited NiO_x reached a maximum efficiency of 7.24% and impressive long-term durability of over 2 months shown in Figure 2d. The results indicated that applications of inorganic ETM and HTM for PSCs would effectively hinder the water and oxygen penetration.[45d] The NiO-based n-i-p heterojunction PSCs with their champion photovoltaic parameters were summarized in Table 1.

Table 1. Summary of photovoltaic characteristics of NiO-based n-i-p heterojunction PSCs. Taken from Ref. 13 with permission of MDPI.

Type	Architecture	V_{OC} (V)	J_{SC} (mA/cm^2)	FF	PCE (%)	Ref.
M	FTO/cp-TiO$_2$/mp-TiO$_2$/mp-NiO/Psk/carbon	0.89	18.2	0.71	11.4	48
M	FTO/cp-TiO$_2$/mp-TiO$_2$/mp-ZrO$_2$/mp-NiO/Psk/carbon	0.965	20.4	0.72	14.2	49
M	FTO/cp-TiO$_2$/mp-TiO$_2$/mp-ZrO$_2$/mp-NiO/Psk/carbon	0.917	21.36	0.76	14.9	50
M	FTO/cp-TiO$_2$/mp-TiO$_2$/mp-Al$_2$O$_3$/mp-NiO/Psk/carbon	0.915	21.62	0.76	15.03	51
P	FTO/cp-TiO$_2$/Psk/NiO$_x$/Ni	0.77	17.88	0.53	7.28	45d

Type "M" or "P" indicates "mesoporous" or "planar" structure respectively. In architecture, "cp" or "mp" means "compact film" or "mesoporous film".
FF, fill factor.

C. Mesoporous NiO-Based p-i-n PSCs

The original application of mesoporous NiO in PSCs was reported by Tian *et al.* and Wang *et al.* nearly at the same time. Tian *et al.* fabricated NiO-based mesoporous PSCs with an architecture of FTO/compact NiO/ mesoporous NiO/sequentially deposited CH$_3$NH$_3$PbI$_3$/PCBM/Al. After controlling the thickness of compact NiO film of 80 nm and that of mesoporous NiO layer of 120 nm, the best device exhibited an efficiency of 1.5% which mainly suffered from low fill factor (FF) and J_{SC} due to high NiO/perovskite interface resistance.[52] Wang *et al.* incorporated a mesoporous NiO as a host to adsorb more amount of perovskite for the improvement of light harvesting.[53] Figure 3a displayed the device structure that was constructed as indium-doped tin oxide (ITO)/compact NiO/ mesoporous NiO/sequentially deposited CH$_3$NH$_3$PbI$_3$/PC$_{61}$BM/BCP/Al. The energy level arrangement, in which the VB position of NiO is closed to that of perovskite and the lowest unoccupied molecular orbital (LUMO) of PC$_{61}$BM is nearly identical with the CB edge of perovskite, is suitable for receiving high output voltage with minimized energy loss for charge separation process between the absorber and selective contacts as seen in Figure 3b. The champion cell delivered a V_{OC} of 1.04 V, a J_{SC} of 13.24 mA/cm^2, and an FF of 0.69, leading to an overall efficiency of

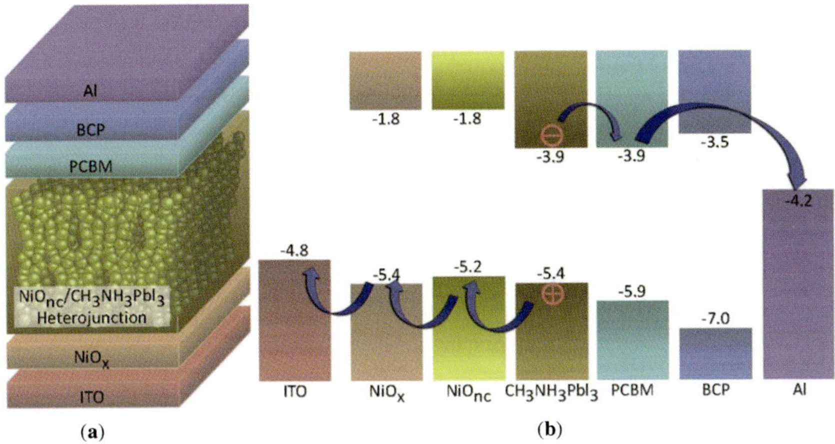

Figure 3. (a) Schematic illustration of NiO-based mesoscopic PSCs device. (b) The energy level of each layer. Taken from Ref. 53 with permission of Nature Publishing Group.

9.51%. Compared with the same group's previously reported planar heterojunction junction (PHJ) device,[54] the implementation of the mesoscopic NiO layer evidently meliorated device performance for increased current response at long wavelength part due to the improvement of light harvesting resulted from a thicker perovskite layer. With elimination of a NiO compact layer, Zhu *et al.* employed an extremely thin mesoporous NiO layer as hole extraction and transport layer for efficient PSC that was consisted of FTO/mesoporous NiO/sequentially deposited $CH_3NH_3PbI_3$/PCBM/Au. A corrugated surface resulted from aggregation of faceted NiO nanocrystals facilitated perovskite formation with good coverage and interconnectivity. By optimizing the thickness of the mesoporous NiO layer of 40 nm, its electron blocking ability was improved to reduce the charge recombination and leakage current. The best performance of NiO-based PSC displayed a V_{OC} of 0.882 V, a J_{SC} of 16.27 mA/cm^2, an FF of 0.635, and a decent PCE of 9.11%.[55] Furthermore, Wang *et al.* introduced a low-temperature sputtered NiO$_x$ film to serve as a pinhole-less compact layer for optimizing charge collection losses in the NiO/perovskite interface. A high-quality NiO$_x$ compact layer formed by sputter deposition has benefits in the charge collection, leading to a remarkable improvement in device performance and reproducibility as compared to the PSCs employing solution-processed NiO_x.[53] A full ITO/sputtered-compact NiO_x/mesoporous NiO/sequentially deposited $CH_3NH_3PbI_3$/PCBM/BCP/Al device was demonstrated, in which the compact NiO_x layer was deposited by radio frequency (RF) magnetron sputtering in pure Ar gas or reactive atmosphere under a mixture of Ar and O_2 gas. By optimizing the oxygen doping ration of 10% and the thickness of mesoporous NiO layer, a decent PCE of 11.6% was obtained.[56]

As revealed from the photoinduced transient absorption (PIA) spectrum, long-lived free carriers in the mesoporous NiO have been identified to prove the charge separation occurred at the NiO/perovskite interface.[45c,53] In addition, replacement of noninjection scaffold is suggested to minimize the interface recombination.[45c] Chen *et al.* reported the implementation of hybrid interfacial layer by incorporating an ultrathin NiO compact layer with a thin mesoporous Al_2O_3 scaffold to fabricate inverted PSCs with a FTO/compact NiO/mesoporous Al_2O_3/$CH_3NH_3PbI_3$/PCBM/BCP/Ag configuration. The hybrid interfacial layer of NiO/mp-Al_2O_3 significantly minimizes light harvesting and interfacial recombination losses, in which thinner compact NiO layer and highly transparent mp-Al_2O_3 were utilized to minimize optical loss. Shunt paths between NiO and PCBM are inhibited by introducing Al_2O_3 nanoparticles on the pinhole of NiO compact layer. Devices using hybrid interfacial layer displayed high FF of 0.72 with a PCE exceeding 13%.[57]

D. Planar NiO-Based p-i-n PSCs

For the demonstration of the first planar NiO-based PSCs, Jeng *et al.* employed a solution-processed NiO_x thin film as an electrode interlayer on the ITO substrate to realize $CH_3NH_3PbI_3$/PCBM PHJ hybrid solar cells. As an evidence of effective photoluminescence (PL) quenching, NiO_x was expected to be a potential electrode for extracting hole carriers. To further overcome the coverage issue, an ultraviolet (UV) ozone (UVO) treatment of the ITO/NiO_x substrate was performed to improve surface wetting capability and resulted in an improvement of the perovskite coverage. The work function of NiO_x thin film was further modified to be 5.4 eV for better matching with the VB edge of $CH_3NH_3PbI_3$. The hybrid cell composed of

ITO/spin-coated NiO$_x$/one-step deposited CH$_3$NH$_3$PbI$_3$/PCBM/BCP/Al was fabricated to deliver an efficiency of 7.8%.[54] Lai *et al.* proposed oxidized-Ni metal film with variant annealing temperature to realize NiO$_x$ compact layer as an HTM. A full device with an ITO/cp-NiO$_x$/one-step deposited CH$_3$NH$_3$PbI$_3$/PCBM/BCP/Al structure was fabricated. By oxidizing Ni metal with an annealing temperature of 450°C, NiO$_x$ film shows a smooth surface morphology that facilitates perovskite interconnectivity with pinhole-less surface coverage, leading to the best PCE of 7.75%.[58] The same group further employed Au:NiO$_x$ to simultaneously function as HTM and conductive electrode by oxidizing this e-beam-sputtered Ni/Au bilayer in O$_2$ atmosphere during annealing process. The bifunctional p-type electrode composed of proper thicknesses of Ni (10 nm) and Au (7 nm) under an annealing temperature of 500°C shows acceptable optical transparent (transmission > 60% in the vis spectrum), good electrical conductivity, and favorable work function matching the VB of perovskite. A TCO-free PSC was fabricated with a device configuration of glass/Au:NiO$_x$/gas-assisted deposition CH$_3$NH$_3$PbI$_3$/C$_{60}$/BCP/Al, delivering a J_{SC} of 13.04 mA/cm^2, a V_{OC} of 1.02 V, an FF of 0.77, and a decent PCE of 10.24%. Improvements on V_{OC} and FF were attributed to the minimized energy loss and superior electrical conductivity, as compared their last work.[59] Subbiah *et al.* used electrodeposited NiO and CuSCN films as HTMs for their FTO/HTM/CH$_3$NH$_3$PbI$_{3-x}$Cl$_x$/PCBM/Al device, in which the perovskite layer was deposited *via* coevaporation of PbCl$_2$ and CH$_3$NH$_3$I.[60] The results indicated that the NiO-based PSCs exhibit superior performance than the CuSCN-based PSCs because of lower series resistance and higher shunt resistance. The NiO-based device with UVO treatment displayed a J_{SC} of 14.2 mA/cm^2, a V_{OC} of 0.786 V, an FF of 0.65, and a PCE of 7.26%.[61] Hu *et al.* fabricated a solution-processed NiO compact layer by spin coating a nickel acetate methoxy-ethanol solution. Similar structure of ITO/NiO/sequentially deposited CH$_3$NH$_3$PbI$_3$/PCBM/Al with UVO treatment was achieved to present device efficiency of 7.6%.[62] To further improve the quality of the NiO, Cui *et al.* exploited the reactive magnetron sputtering method to make a compact NiO film for a cell structure of FTO/sputtered NiO/solvent-engineering CH$_3$NH$_3$PbI$_3$/PCBM/BCP/Au. The solvent engineering method was introduced to form a homogeneous surface coverage and uniform interlayer morphology of perovskite absorber onto the compact layer. With a 240-nm-thick perovskite and effective NiO electrode, the as-fabricated device exhibited an increasing PCE of 9.84%.[63] The versatility of fabrication including one-step spin coating deposition, two-step sequential deposition, and vapor deposition were extensively used to prepare perovskite film, among which one-step spin-coated perovskite surface exhibited many pinholes and non-homogeneous coverage. To further improve the surface coverage and uniformity of $MAPbI_{3-x}Cl_x$, Li *et al.* developed a preheating-assisted deposition method to produce a pinhole-less film, delivering a PCE of 6.4% for the inverted architecture.[64] Similarly, Xiao *et al.* applied spin-coated NiO_x as an interlayer in inverted structure of ITO/NiO_x/$CH_3NH_3PbI_3$/PCBM/C_{60}/BCP/Al and compared its performance with the hole acceptor of VO_x and MoO_x as counterparts. A homogeneous and uniform perovskite photoactive absorber with micron-sized grains was prepared using gas-assisted deposition method.[65] As demonstrated from the steady-state PL responses of perovskite deposited on NiO_x, VO_x, and MoO_x, similar magnitude of PL quenching was observed, implying these inorganic p-type semiconductors as efficient HTMs. The fabricated NiO_x-based PSC displayed a compatible PCE of $10.6 \pm 0.6\%$ compared to $11.7 \pm 0.5\%$ of VO_x-based device, while MoO_x-based solar cell exhibited an inferior PCE of $5.9 \pm 0.5\%$ due to the mismatched energy levels.[66]

On the whole, NiO-based PSCs generally exhibited a lower FF than the PEDOT-based devices although they displayed a higher V_{OC}. PSC applying hybrid NiO_x/PEDOT HTM was further proposed by Park *et al.* to improve FF. With additive of PEDOT covering on NiO_x-coated substrate, the NiO_x surface morphology was modified to be rough, which benefits increased interfaces of NiO_x/$CH_3NH_3PbI_3$ heterojunction and reduces interface resistance. A planar p-i-n PSC of ITO/hybrid NiO_x-PEDOT/$CH_3NH_3PbI_3$/PCBM/Ag with high conversion efficiency of 15.1% was presented by using PEDOT treatment of 1.0% (v/v). As proven by the PL spectra and impedance spectroscopy, PL quenching and interface charge transfer resistance (R_{CT}) at the NiO_x/perovskite interface are significantly improved with the PEDOT treatment. The experimental results also exhibited neglected hysteresis of the fabricated inverted PSC device.[67] Li *et al.* used an extremely thin NiO_x film of 10 nm as a hole extraction layer for a device composed of ITO/spin-coated

$NiO_x/CH_3NH_3PbI_3/C_{60}/BCP/Ag$. A high-quality perovskite layer was deposited by fast deposited crystallization[68] to form a compact surface coverage and flat interlayer morphology, and the NiO-based planar PSCs delivered a V_{OC} of 994 ± 15 mV, a J_{SC} of 20.4 ± 0.7 mA/cm^2, an FF of 0.669 ± 0.022, and a PCE of 13.6 ± 0.6%. It is surprising that an HTM-free PSC was further demonstrated to exhibit superior performance compared to NiO_x-based PSCs. In the HTM-free PSCs, perovskite simultaneously functions as the light absorber and hole transporter.[69] With a UVO treatment of ITO substrate, the work function of ITO was improved to a higher value of 5.2 eV that reduced the energy barrier between ITO and perovskite layer. Moreover, UVO treatment effectively removes the residual organic species to lower surface energy, which benefits the deposition of perovskite layer. On the other hand, the perovskite layer deposited on the bare ITO substrate with a nano-sized rough surface morphology improved the perovskite/ITO heterojunction interface. Eventually, the HTM-free PSCs showed a higher J_{SC} of 21.7 ± 0.5 mA/cm^2, a higher FF of 0.718 ± 0.015, and a promising PCE of 15.0 ± 0.5%.[70] The time-dependent capacitive current significantly affected the output current, which resulted in the J-V hysteresis.[71] As a result, unraveled origin of hysteresis in $NiO/MAPbI_3/PCBM$ inverted structure was elucidated by Kim *et al.* by using impedance spectroscopy to measure frequency-dependent capacitance. The capacitance response in intermediate frequency was attributed to the electronic polarization in the bulk perovskite, while that in low frequency was dominated by the interfacial charge accumulation at the $MAPbI_3$/electrode interface. The capacitance at low frequency assigned as electrode polarization was measured to ~10^{-7} F/cm^2 for $NiO/MAPbI_3/PCBM$ heterojunction, which was below the typical capacitance of 10^{-5}–10^{-6} F/cm^2. In addition, the characteristic frequency, at which the electrode polarization is transformed to the dipole polarization, for $NiO/MAPbI_3/PCBM$ heterojunction shifted to higher frequency of ~10^2 Hz. These evidences indicate that the reduced capacitance of electrode polarization is much faster than the timescale for J-V hysteresis; thus, the hysteresis was apparently eliminated for the inverted $NiO/MAPbI_3/PCBM$ device.[72]

Doping NiO with high conductive metal ions or modifying $NiO/$perovskite interface heterojunction made significant improvements on

NiO-based PSCs. Kim *et al.* demonstrated a high-efficiency planar PSC by using solution-processed copper (Cu)-doped NiO_x ($Cu:NiO_x$) as an HTM. With copper acetate doping in nickel acetate as precursor, a Cu-doped NiO_x was fabricated *via* the conventional solgel route with a sintering process at 500°C to achieve high crystallinity. The device structure of $ITO/Cu:NiO_x/CH_3NH_3PbI_3/PC_{61}BM/C_{60}/Ag$ was demonstrated with a high PCE of 15.4%, and the improvement was attributed to the enhanced electrical conductivity and charge extraction capability of $Cu:NiO_x$ film, and favorable perovskite crystallization on $Cu:NiO_x$ film. Moreover, the application of wider band gap Br-doped perovskite solar absorber in $Cu:NiO_x$-based PSCs was further demonstrated to exhibit a high voltage output of 1.13 and 1.16 V for $MAPb(I_{0.8}Br_{0.2})_3$ and $MAPb(I_{0.6}Br_{0.4})_3$ devices, respectively, because of less potential loss between work function of NiO_x and VB of Br-contained perovskite.[73] In the following progress, the same group further modified the preparation method to fabricate $Cu:NiO_x$ by a low-temperature combustion process that was prepared by doping copper nitrate trihydrate into nickel nitrate hexahydrate and acetylacetone as precursor. A highly crystalline $Cu:NiO_x$ *via* combustion process at 150°C exhibited similar work function of -5.3 eV compared to that of solgel-processed $Cu:NiO_x$. Moreover, combustion-processed $Cu:NiO_x$ exhibited higher electrical conductivity than that of high-temperature-sintered film by almost twofolds. PSCs fabricated with the configuration of $ITO/Cu:NiO_x/CH_3NH_3PbI_3/C_{60}/Bis-C_{60}/Ag$ displayed a recorded PCE of 17.74%.[74] The effect of UVO irradiation upon Cu:NiO film prepared by copper-doped nick acetate was explored to work as an efficient HTM for inverted PSCs. UVO-treated Cu:NiO film with annealed process exhibited superior electrical conductivity because of fully oxidation in Cu:NiO film. Such treatment also improved the crystallization of perovskite with better coverage due to better wettability and higher roughness as compared to pristine one. Moreover, the underlying Cu:NiO layer with UVO irradiation delivered a deeper work function of 5.48 eV that aligned well with the VB of perovskite. The NiO-based device made of $ITO/UVO-treated Cu:NiO/CH_3NH_3PbI_3/PC_{60}BM/Al$ showed an enhanced efficiency of 12.2% compared to the untreated Cu:NiO-based device.[75]

Park *et al.* modified the NiO/perovskite interface heterojunction by employing the pulse laser deposition (PLD) method to fabricate a well-ordered nanostructured NiO thin film with a good optical transparency and preferred orientation of (111). The (111)-oriented NiO showed a lower resistivity and sheet resistance compared to other growth directions, which have the benefit of effectively extracting hole carrier and preventing recombination. Also, the well-ordered nanostructured NiO provides more contact interface with perovskite to facilitate charge extraction. The full device was constructed as ITO/PLD-NiO/CH$_3$NH$_3$PbI$_3$/PCBM/LiF/Al with a dense perovskite prepared by solvent engineering technique. By introducing oxygen partial pressure of 200 mTorr and controlling deposition time, a nanostructured NiO film with controlled thickness was made for high-efficiency planar p-i-n heterojunction PSC. The best performing device exhibited a J_{SC} of 20.2 mA/cm^2, a V_{OC} of 1.06 V, and a remarkable efficiency of 17.3% with a very high FF of 0.813.[76] Moreover, Seo *et al.* further modified the NiO/perovskite interface heterojunction by growing a conformal, dense, and ultrathin NiO film with atomic layer deposition (ALD) method, of which the thickness was precisely controlled to be comparable with the Debye length. When the thickness of NiO film approached to the Debye length, the boundary space charge distribution overlapped within the NiO film (Figure 4c), leading to an increase of hole concentration and a favorable work function of ~5.0 eV. The inverted device of ITO/ALD-NiO/CH$_3$NH$_3$PbI$_3$/PCBM/Ag delivered an efficiency of 11.93%. Moreover, the electrical conductivity of NiO film was further improved by using annealing process to reduce surface hydroxylated NiOOH and C-N related defects. Eventually, the as-fabricated PSCs with

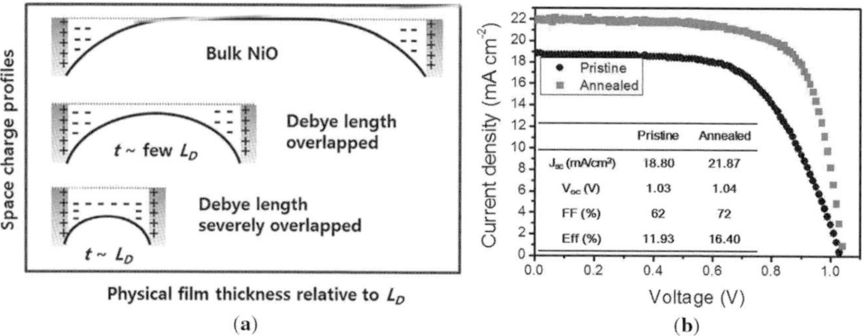

Figure 4. (a) Schematic images of space charge profile relative to the NiO film thickness. (b) *J-V* curve for the PSC using ALD-NiO with a thickness of 7.5 nm. Taken from Ref. 77 with permission of the Royal Society of Chemistry.

annealed NiO film exhibited a negligible hysteresis and a promising PCE of 16.4% as shown in Figure 4b.[77]

To further simplify the deposition procedure, Yin *et al.* employed solution-processed NiO_x film to fabricate an inverted planar PSC that was composed of $FTO/NiO_x/CH_3NH_3PbI_3/PCBM/Ag$. The thickness of NiO_x film significantly influences the photovoltaic performance, hysteresis effect, and air storage stability of the as-fabricated devices. The thickness-dependent roughness would have impact on the charge collection. With a suitable thickness of 90 nm and nanostructure roughness of 3.835 nm for NiO_x compact layer, a hysteresis-free NiO_x-based PSC was fabricated with a PCE of 14.42%.[78] The same group further employed simplified process to presynthesize high-quality NiO_x nanoparticles solution. With the replacement of FTO substrate with ITO substrate, a reduced NiO_x film of 50 nm presented fully converge on the ITO substrate with reduced series resistance. The solar cell with the same configuration using low-temperature solution-processed NiO_x film as a p-type contact reached superior efficiency of 16.47%. The low-temperature process allows the NiO_x film to be deposited on the flexible ITO-PEN substrate, and an acceptable PCE of 13.43% was delivered.[79] Moreover, spin coating NiO_x film *via* solution process was further applied to flexible PET substrate. Zhang *et al.* demonstrated flexible PSCs employing a room temperature formation of NiO_x film to display an impressive PCE of 14.53% for the first time. A structure of $PET/ITO/NiO_x/CH_3NH_3PbI_3/C_{60}/Bis-C_{60}/Ag$ was achieved, in which the formed NiO_x film presents a feature of flawless nanostructured morphology that improves the hole extraction, and reduces the interfacial recombination and monomolecular Shockley–Read–Hall recombination of as-fabricated PSCs. Hysteresis-free with good air and mechanical stability of flexible NiO_x-based PSCs were thus achieved.[80]

It is worth noting that the abovementioned structures usually use PCBM as an ETM, whose stability and cost are the issues of concerns. Replacing organic charge extraction layers with inorganic materials would improve the devices' stability and provide versatile choices for materials selection and device design. For the fabrication of durable and cost-effective PSCs, You *et al.* reported planar PSCs with all solution-processed metal oxide charge transport layers, whose device structure consisted of ITO/NiO$_x$/CH$_3$NH$_3$PbI$_3$/ZnO/Al as shown in Figure 5a. Favorable band alignment between NiO$_x$ film and perovskite material, and improved crystallinity of perovskite film on NiO$_x$ lead to cells with higher V_{OC} as compared to cells using PEDOT:PSS. On the other hand, ZnO film possesses strong hole-blocking effect originating from its deep VB, resulting in a higher J_{SC} for the ZnO-based PSCs. The all metal-oxide-based PSCs achieved a V_{OC} of 1.01 V, J_{SC} of 21.0 mA/cm^2, and an FF of 0.76, leading to a remarkable efficiency of 16.1%. With replacement of aforementioned PCBM with ZnO, a significant enhancement in stability of un-encapsulation cell was achieved under ambient environment for 60 days as shown in Figure 5b.[45a] By heavily doped metal oxide semiconductor as charge transport layer in planar PSCs, device performance of large active area (>1 cm^2) with high-certified efficiency (>15%) and superior stability under light soaking were further demonstrated. Chen et $al.$ employed heavily p-doped (p$^+$) NiMgLiO and n-doped (n$^+$) Ti(Nb)O$_x$ as HTM and ETM, respectively, in inverted planar PSCs to improve charge transport capability and minimize resistive losses. As confirmed by transient photocurrent and photovoltage decays measurements, faster charge transport through the heavily doped charge carrier extraction layers facilitate the charge

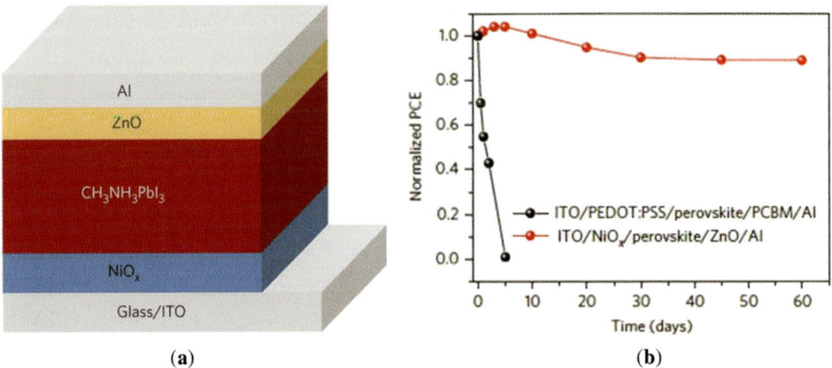

Figure 5 (**a**) Scheme of PSCs employing all metal oxide charge transport layers. (**b**) Stability of the device in (a) kept in ambient environment without encapsulation. Taken from Ref. 45a with permission of Nature Publishing Group.

extraction and prevent charge accumulation at the interface of HTM/$CH_3NH_3PbI_3$/ETM heterojunction. A remarkable efficiency of 18.3% is obtained with a very high FF of 0.83 for small aperture area of 0.09 cm^2. The encapsulated PSC achieved promising stability under light soaking for 1000 h. The PSCs employing highly doped metal oxide charge extraction layers were fabricated with active area > 1 cm^2, and the cell showed a remarkable efficiency of 16.2%.[45b] Based on the inverted configuration, all metal-oxide-based PSCs display a small hysteresis with different scan directions. The NiO-based p-i-n heterojunction PSCs with their champion photovoltaic parameters were summarized in Table 2.

E. Charge Separation, Dynamics, Kinetics and Chemical Reaction at Perovskite/NiO Interface

With a remarkable swift progress of photovoltaic performance for the NiO-based inverted PSC, fundamental understandings have been conducted much later with limited investigations. It was until very recent that several experimental results deciphered the underlying principle and mechanism for such efficient power conversion system. In general, PL quenching effect of the perovskite/NiO junction has been extensively demonstrated as a proof for efficient photoexcited carrier extraction. The PL spectra of perovskite coated on mesoporous Al_2O_3 or mesoporous NiO are shown in Figure 6a.[81] It is noted that a scaffold layer of mesoporous Al_2O_3 does not quench the PL due to its wide band gap. To quantitatively analyze the PL quenching of both films, the steady-state PL spectra with an excitation wavelength at 450 nm were recorded (dashed curves in Figure 6a). As the absorption observed at this wavelength in the UV–vis is very similar, the results indicated that PL intensity of the perovskite coated on the NiO film was quenched more significantly than that deposited on the Al_2O_3 film. To understand the mechanism of charge transport at the perovskite/NiO interface, the direct evidence of charge separation was firstly examined by PIA method. From the PIA spectrum shown in Figure 6b, perovskite absorption edge bleaching was clearly recognizable in the spectral region below 800 nm, together with a broad free-carrier absorption band in the spectral wavelength of 800–1500 nm upon

Table 2. Summary of photovoltaic characteristics of NiO-based p-i-n heterojunction PSCs. Taken from Ref. 13 with permission of MDPI.

Type	Architecture	V_{OC} (V)	J_{SC} (mA/cm^2)	FF	PCE (%)	Ref.
M	FTO/cp-NiO/mp-NiO/Psk/PCBM/Al	0.83	4.9	0.35	1.5	[52]
M	ITO/cp-NiO/mp-NiO/Psk/PCBM/BCP/Al	1.04	13.24	0.69	9.51	[53]
M	FTO/mp-NiO(solgel)/PsK/PCBM/Au	0.882	16.27	0.635	9.11	[55]
M	ITO/sputtered-NiO/mp-NiO/Psk/PCBM/BCP/Al	0.96	19.8	0.61	11.6	[56]
M	FTO/NiO/meso-Al$_2$O$_3$/Psk/PCBM/BCP/Ag	1.04	18.0	0.72	13.5	[57]
P	ITO/NiO$_x$/Psk/PCBM/BCP/Al	0.92	12.43	0.68	7.8	[54]
P	ITO/NiO$_x$ (Ni-oxidized)/Psk/PCBM/BCP/Al	0.901	13.16	0.65	7.75	[58]
P	(TCO-free) Au:NiO$_x$/Psk/C$_{60}$/BCP/Al	1.02	13.04	0.77	10.24	[59]
P	FTO/NiO/Psk/PCBM/Al	0.786	14.2	0.65	7.26	[61]
P	ITO/NiO/Psk/PCBM/Al	1.05	15.4	0.48	7.6	[62]
P	FTO/sputtered-NiO/Psk/PCBM/BCP/Au	1.10	15.17	0.59	9.84	[63]
P	ITO/NiO$_x$/Psk/PCBM/Au	0.798 ± 0.03	18.23 ± 0.8	0.47 ± 0.03	6.4 ± 0.5	[64]
P	ITO/NiO$_x$/Psk/PCBM/C$_{60}$/BCP/Al	1.06 ± 0.05	18.0 ± 2.2	0.56 ± 0.08	10.6 ± 0.6	[66]
p	FTO/NiO$_x$/Psk/PCBM/Ag	1.09	17.93	0.74	14.42	[78]
P	ITO/NiO$_x$/Psk/PCBM/Ag	1.07	20.58	0.75	16.47	[79]
P	ITO/Cu:NiO$_x$(solgel)/Psk/PCBM/C$_{60}$/Ag	1.11 ± 0.01	18.75 ± 0.42	0.72 ± 0.01	14.98 ± 0.33	[73]

(Continued)

Table 2. (*Continued*)

Type	Architecture	V_{OC} (V)	J_{SC} (mA/cm^2)	FF	PCE (%)	Ref.
P	ITO/Cu:NiO$_x$(combustion)/ PsK/C$_{60}$/Bis-C$_{60}$/Ag	1.05	22.23	0.76	17.74	[74]
P	ITO/UVO Cu:NiO (annealed)/Psk/PC$_{60}$BM/Al	1.00 ± 0.02	16.1 ± 1	0.67 ± 0.04	12.2	[75]
P	ITO/PLD-NiO/Psk/PCBM/LiF/Al	1.06	20.2	0.813	17.3	[76]
P	ITO/ALD-NiO (annealed)/Psk/PCBM/Ag	1.04	21.9	0.72	16.4	[77]
P	PET/ITO/NiO$_x$/ Psk/C$_{60}$/Bis-C$_{60}$/Ag	0.97 ± 0.01	20.55 ± 0.71	0.68 ± 0.03	13.33 ± 0.78	[80]
P	ITO/hybrid NiO$_x$-PEDOT/Psk/PCBM/Ag	1.02 ± 0.006	19.4 ± 0.3	0.70 ± 0.016	13.9 ± 0.4	[67]
P	ITO/NiO$_x$/Psk/C$_{60}$/BCP/Ag	0.994 ± 0.015	20.4 ± 0.7	0.669 ± 0.022	13.6 ± 0.6	[70]
P	ITO/NiO$_x$/Psk/ZnO/Al	1.01	21.0	0.76	16.1%	[45a]
P	FTO/NiMgLiO/Psk/PCBM/Ti(Nb)O$_x$/Ag (active are>1 cm^2)	1.072	20.21	0.748	16.2%	[45b]

Type "M" or "P" indicates "mesoporous" or "planar" structure respectively. In architecture, "cp" or "mp" means "compact film" or "mesoporous film".

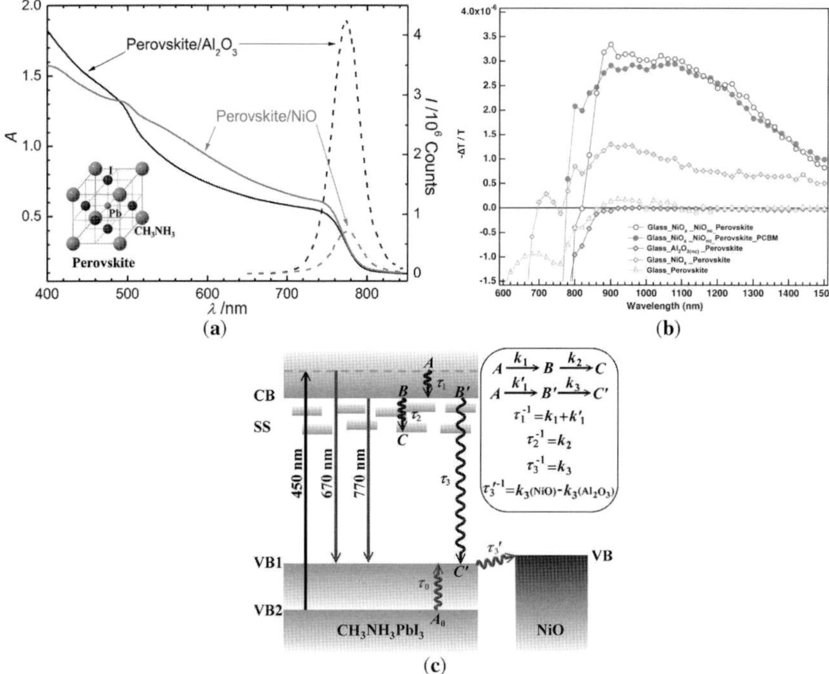

Figure 6. (a) Absorbance (solid line) and photoluminescence (dash line) spectra of perovskite/Al$_2$O$_3$ (black) and perovskite/NiO (gray) films. Taken from Ref. 81 with permission of Wiley-VCH. (b) The PIA spectra of perovskite in contact with various materials. Taken from Ref. 53 with permission of Nature Publishing Group. (c) Mechanism of excitonic relaxation showing radiative and nonradiative processes observed for perovskite coated on mesoporous NiO semiconductor films. A kinetic model for fitting the PL transients was shown on the top right box. Taken from Ref. 81 with permission of Wiley-VCH.

photoexcitation of the perovskite/NiO film.[45c,53] Such free-carrier absorption feature implies that long-lived charge-separation state occurred in the perovskite/NiO junction. On the contrary, when perovskite is in contact with noninjecting wide band gap materials (Al$_2$O$_3$ and glass), no such IR transient absorption feature is present.

Hsu *et al.* used femtosecond photoluminescence up-conversion spectroscopy to further realize the carrier relaxation mechanism in mesoporous

NiO/perovskite together with mesoporous Al_2O_3/perovskite for comparison.[81] Figure 6c schematically illustrated the relaxation mechanism, in which the excitation at 450 nm invokes a transition from the VB_2 state to the hot CB state, and the probe at 670 and 770 nm correspond to the transitions of hot CB/VB_1 and cold CB/VB_1, respectively. The transient followed the kinetic model containing two sequential relaxations in parallel: (R1) A→B→C and (R2) A→B'→C'. Component A represents the hot electrons relaxation of $CH_3NH_3PbI_3$ in the CB state. The carriers thermalized to CB edge subsequently produce components B and B'. Components B and B' represent the relaxation to the surface state (SS; C) and the internal recombination relaxation to the VB1 state (C'), respectively. In experimental measurements, the temporal emission of PL transients were probed in the spectral region of 670–810 nm, in which individual PL transient was deconvoluted into one rapid decay (0.2–0.5 ps) of component A (contribution from hot carrier relaxation) and two slow decays of components B (20–50 ps for SS relaxation) and B' (0.5–1.2 ns for recombination). The rate of the decay of component B is similar for both films, whereas the decay coefficient of component B' of the Al_2O_3 film is significantly greater than those of the NiO film. This observation is correlated to a decreased relation time of τ_3 of the NiO film and contributed to hole transport at the NiO/perovskite interface. The hole extraction time from perovskite by the NiO is estimated to be around 5 ns at the PL maximum of 770 nm. On the other hand, the similar relaxation times of τ_2 observed for both Al_2O_3 and NiO films indicated that it is the intrinsic relaxation process that occurred on the surface of the perovskite. For the amplitude of component B, the intensity in the perovskite/NiO film is greater than that of perovskite/Al_2O_3, which indicated that quenching by the nonemissive SS is more prevailing. The observed quenching of the steady-state PL intensity of perovskite, shown in Figure 6a, might be attributed to the surface-state relaxation of perovskite, which is more significant for the NiO film than for the Al_2O_3 film.

Corani *et al.* also specified the timescale of hole injection as well as recombination dynamics from the photoexcited perovskite to the nanocrystalline NiO ($NiO_{(nc)}$) electrode using time-resolved terahertz spectroscopy (TRTS). TRTS was a useful tool in measuring photoconductivity and carrier mobility of materials with a noncontact manner to minimize the errors in measuring its electrical properties. Further detail can be found in another chapter of this theme book. Early timescale of carrier dynamics in terms of photoconductivity for four samples were compared and displayed in Figure 7a, including quartz/MAPbI$_3$, quartz/NiO$_{(nc)}$/MAPbI$_3$, quartz/TiO$_2$/MAPbI$_3$, and quartz/ZrO$_2$/MAPbI$_3$, and there is no expected charge injection into ZrO$_2$ because of its high band gap of 5.34 eV. When MAPbI$_3$ is attached to NiO$_{(nc)}$, the mobility was reduced by half compared to neat perovskite and perovskite/ZrO$_2$, and the rise time of the signal was on the sub-picosecond timescale. These results indicated that an ultrafast injection of holes occurred from MAPbI$_3$ to NiO$_{(nc)}$ driven by the favorable band alignment between perovskite and NiO$_{(nc)}$. The reduction in the photoconductivity is a result of the disappearance of highly mobile hole from the perovskite. Because of the balanced charge transport and the very similar diffusion lengths of hole or electron carriers in perovskite, similar photoconductivity was observed for the perovskite/TiO$_2$ sample. The photoconductivity kinetics of MAPbI$_3$/NiO$_{(nc)}$ under variant excitation intensities were further conducted to determine the timescale of population dependent recombination as shown in Figure 7b. An apparent decay was presented under all excitation intensities owing to the recombination between holes injected into NiO$_{(nc)}$ and mobile electrons in the

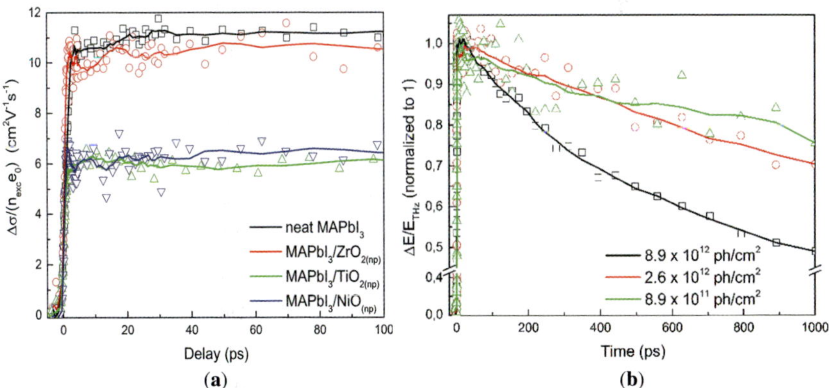

Figure 7. (**a**) Early time THz photoconductivity kinetics of MAPbI$_3$/NiO heterojunction. (**b**) THz photoconductivity kinetics of MAPbI$_3$/NiO heterojunction pumped at three different intensities. Taken from Ref. 82 with permission of the American Chemical Society.

perovskite. From Figure 7b, the recombination time was estimated to be hundreds of picoseconds to a few nanoseconds and is highly population dependent. Briefly, when the hole carriers rapidly injected into the $NiO_{(nc)}$, hole carriers are slowing down due to the low conductivity of $NiO_{(nc)}$. Thus, the sum of photoconductivity is reduced and recombinations between electrons (in perovskite) and holes (in NiO) begin to take place. This recombination process competes with charge collection, which has great influence on the device performance. From the earlier discussions, promote higher hole mobility by doping NiO or improving the perovskite/NiO heterojunction interface is desirable for the aim of preventing back recombination.[82]

To further realize the origin of highly efficient carrier transport for NiO/perovskite heterojunction, Lin *et al.* conducted X-ray photoelectron spectra (XPS) to analyze the chemical environment of the NiO/perovskite interface.[83] Figure 8 identified the chemical states of Pb 4f, I 3d, N 1s, and O 1s at various interfaces (ITO/PbI_2, NiO/PbI_2, and NiO/perovskite). In Figure 8a-top, the Pb $4f_{7/2}$ spectrum decoupled into three pecks at BE 138.2, 137.8, and 136.7 eV that are respectively assigned to PbI_2, PbO, and Pb for ITO/$NiO_{(nc)}$/PbI_2. On the other hand, the Pb $4f_{7/2}$ spectrum of ITO/$NiO_{(nc)}$/$MAPbI_3$ was composed of three signals at BE 137.9, 137.5, and 136.7 eV, which are respectively corresponding to $MAPbI_3$, $CH_3NH_3PbI_{3-2\delta}O_\delta$, and Pb. The characteristic BE at 138.2 eV for PbI_2 shifted to 137.9 eV for $MAPbI_3$ (Figure 8a-middle) after the formation of $MAPbI_3$ by using sequential deposition. As seen in Figure 8b-top, the ITO/$NiO_{(nc)}$/PbI_2 spectrum was deconvoluted into two clear features at BE 619.9 and 618.9 eV, which were respectively assigned to I_2 and PbI_2. After the ITO/$NiO_{(nc)}$/PbI_2 sample was immersed in ammonium iodide (MAI) solution, the I 3d signals were composed of three discernible features at 619.9 eV for I_2, 619.6eV for $CH_3NH_3PbI_{3-2\delta}O_\delta$, and 618.6 eV for $MAPbI_3$.

For the ITO/$NiO_{(nc)}$ sample (Figure 8c-bottom), two characteristic features of $NiO_{(nc)}$ at 854.1 and 855.7 eV were attributed to NiO and Ni_2O_3 states, respectively. The shoulder peak located at 861 eV was assigned as the satellite feature. After deposition of PbI_2 on $NiO_{(nc)}$ (Figure 8c-top), an additional feature positioned at smaller BE of 853.0 eV was attributed to the metallic Ni. Compared to Figure 8a-top, the reduction

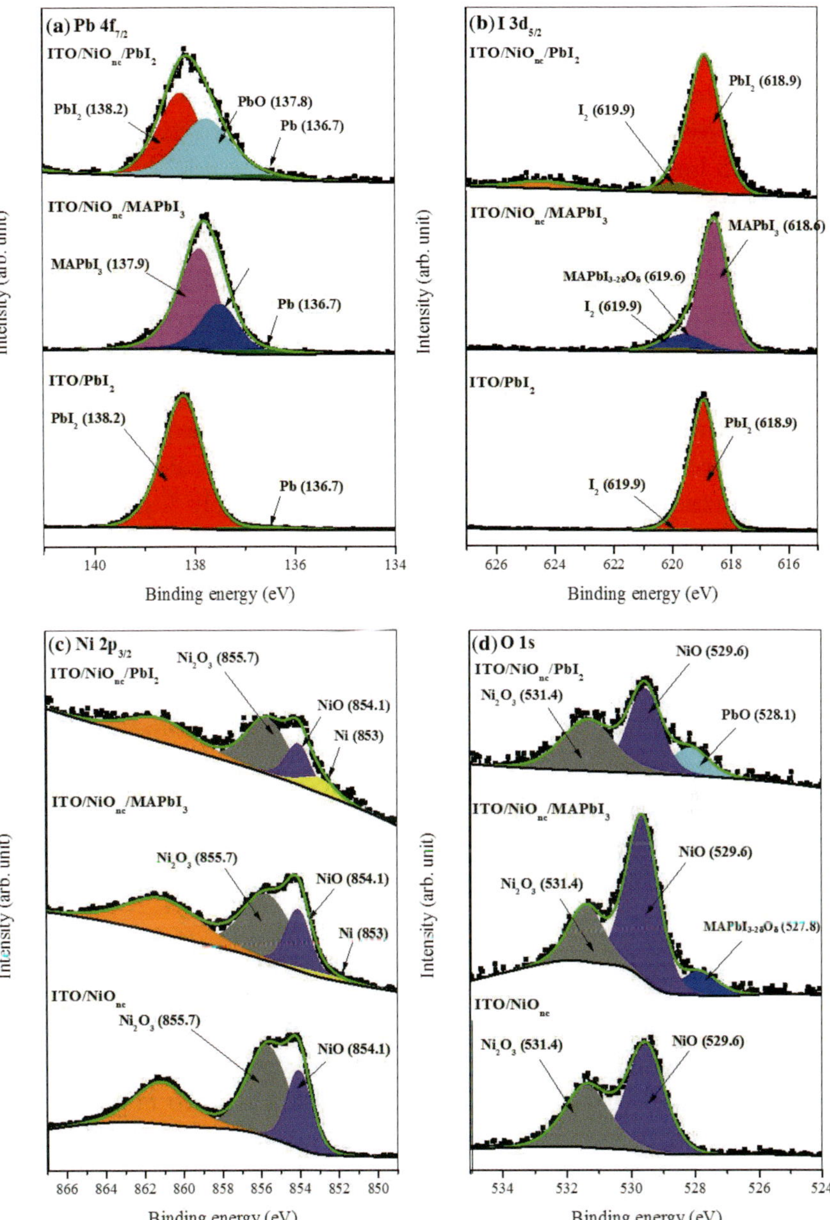

Figure 8. XPS spectra of (**a**) Pb 4f$_{7/2}$, (**b**) I 3d$_{5/2}$, (**c**) Ni 2p$_{3/2}$, and (**d**) O 1s of ITO/NiO$_{(nc)}$/PbI$_2$, ITO/NiO$_{(nc)}$/MAPbI$_3$, together with ITO/NiO$_{(nc)}$ and ITO/PbI$_2$ sample for comparison. Taken from Ref. 83 permission of Wiley-VCH.

of NiO takes place when interacting with PbI_2 and produces the metallic Ni. At the same time, the oxidation of PbI_2 occurred when reacting with NiO to yield the PbO. The metallic Ni may benefit hole extraction from light absorber to the $NiO_{(nc)}$ due to decrease of the $NiO_{(nc)}$/perovskite resistance. Figure 8d displayed the O 1s XPS spectra of ITO/$NiO_{(nc)}$, ITO/$NiO_{(nc)}$/PbI_2, and ITO/$NiO_{(nc)}$/$MAPbI_3$. The oxygen signals corresponding to Ni_2O_3, NiO, and PbO were clearly observed at 531.4, 529.6, and 528.1 eV after PbI_2 deposition on $NiO_{(nc)}$ shown in Figure 8d-top, the former two signals agreed well with that of ITO/$NiO_{(nc)}$ (Figure 8d-bottom). As a result, the formation of metallic nickel and PbO indicates an interfacial redox reaction when $NiO_{(nc)}$ was processed with PbI_2. Through pouring an MAI solution on ITO/$NiO_{(nc)}$/PbI_2 to form the perovskite, PbO became involved to form $CH_3NH_3PbI_{3-2\delta}O_\delta$ at 527.8 eV. Consequently, the PbI_2 was oxidized firstly to PbO and subsequently formed as hole-dopant in $CH_3NH_3PbI_{3-2\delta}O_\delta$ at the interface after the sequential deposition. The generation of hole-doping $CH_3NH_3PbI_{3-2\delta}O_\delta$ facilitates the carrier transport, leading to an improvement of the photovoltaic performance. Based on calculations of electronic structure, the formation of $CH_3NH_3PbI_{3-2\delta}O_\delta$ at the $NiO_{(nc)}$/$MAPbI_3$ heterojunction caused an upper shifted VB matching well with the VB of $NiO_{(nc)}$, which was favorable for hole transport from $MAPbI_3$ to $NiO_{(nc)}$. This interfacial chemical redox reaction plays a key role in achieving a large J_{SC} and a good FF performance of a PSC that should be taken into consideration in designing highly efficient PSC devices.

III. Copper-Based Inorganic HTMs for PSCs

Copper-based inorganic semiconducting hole materials have shown promises for their use in OPVs, dye-sensitized and quantum dot-sensitized solar cells. These wide band gap materials can be deposited with solution process for high conductivity, low-energy budget process, and good transparency.

A. CuSCN as HTM

Copper thiocyanate (CuSCN) serves as a universal hole-transport interlayer material for numerous optoelectronic applications, including transparent thin-film transistors, organic and PSCs, and organic light-emitting diodes. CuSCN has intrinsic hole-transport (p-type) characteristics with a large band gap (>3.5 eV) and good optical transparency throughout the visible to near-infrared spectrum. This material has been used both in n-i-p and p-i-n heterojunction for PSCs.[84]

1. *Mesoporous CuSCN-Based n-i-p PSCs*

CuSCN was first employed by Ito *et al.* into mesoporous PSC with a structure composed of FTO/compact TiO_2/mesoporous TiO_2/$CH_3NH_3PbI_3$/ CuSCN/Au. The $CH_3NH_3PbI_3$ light absorber was fabricated by one-step spin coating[8] and its thickness is controlled by hot-air drying during spin coating. With the introduction of doctor-blade-processed CuSCN as an HTM and a controlled thickness of perovskite, the first CuSCN-based PSCs achieved a PCE of 4.85%. The experiment also indicates that the CuSCN films mitigate photovoltaic degradation of moisture-sensitive perovskite material.[85] The same group further inserted Sb_2S_3 at the interface between mesoporous TiO_2 and perovskite layer to facilitate the crystallization of perovskite formation and serve as a passivation layer on TiO_2 to avoid decomposition of perovskite crystal upon exposure to light. The device showed an enhancement in performance to display a PCE of 5.12% and exhibited stability against light soaking without encapsulation.[86] A remarkable improvement under the same configuration was further achieved by the sequential deposition method[87] and preheating the substrate before PbI_2 deposition. The efficiency was enhanced to 10.51% because of a pinhole-less perovskite layer that limited the diffusion of CuSCN into $CH_3NH_3PbI_3$ active layer.[88] The thickness of the perovskite film, perovskite capping layer, and surface morphology were further controlled with the sequential deposition method. With the application of inorganic p-type CuSCN made by doctor blading, the CuSCN-based solar cell displayed a promising performance with a J_{SC} of 19.7 mA/cm^2, a V_{OC} of 1016 mV, an FF of 0.62, and a PCE of 12.4%.[89] With addition of a small amount of MAI and DMSO in the PbI_2-DMF precursor, the light absorber of perovskite was made by the modified sequential two-step spin coating method to form a pinhole-less film. A pinhole-less perovskite film effectively inhibited possible short-circuiting contact between TiO_2 and CuSCN, resulting in enhancement on photovoltaic performance. A mesoporous PSC with the same structure of FTO/compact TiO$_2$/mesoporous TiO$_2$/CH$_3$NH$_3$PbI$_3$/CuSCN/Au displayed a V_{OC} of 0.96 V, a J_{SC} of 18.23 mA/cm^2, a FF of 0.68, and a PCE of 11.96%. A planar structure is fabricated for comparison to show a lower FF of 0.4, leading to a lower efficiency of 7.19%.[90]

2. *Planar CuSCN-Based n-i-p PSCs*

For the implementation of CuSCN-based planar architecture, Chavhan *et al.* fabricated a solar cell composed of FTO/compact TiO$_2$/CH$_3$NH$_3$PbI$_{3-x}$Cl$_x$/CuSCN/Au. An organometal halide perovskite layer was deposited by spin coating with a thermal annealing process.[91] The champion solar cell was exhibited for an annealing temperature of 110°C, leading to 6.4% PCE. A lower V_{OC} of 727 mV was obtained and it was attributed to nonuniformity and poor surface coverage of CH$_3$NH$_3$PbI$_{3-x}$Cl$_x$ on the compact layer.[92]

3. *Planar CuSCN-Based p-i-n PSCs*

Subbiah *et al.* reported the use of CuSCN for the planar OPV-like structure consisting of FTO/electrodeposited CuSCN/coevaporation CH$_3$NH$_3$PbI$_{3-x}$Cl$_x$/PCBM/Ag with a maximum efficiency of 3.8%. When the perovskite film deposited on the CuSCN film, a considerable PL quenching was observed to prove its effective charge extraction capability. However, without optimizing CuSCN film thickness, high series resistance and low shunt resistance result in poor photovoltaic performance for the CuSCN-based PSCs.[61] Ye *et al.* used electrodeposited CuSCN film as HTM to fabricate inverted planar PSCs based on the structure of ITO/CuSCN/CH$_3$NH$_3$PbI$_3$/C$_{60}$/BCP/Ag. By adopting one-step fast deposition-crystallization (FDC) method,[68] a high-quality perovskite layer was formed on top of CuSCN film. Lower interface resistance between perovskite layer and CuSCN leads to pronounced improvements in J_{SC} and FF from 18.9 ± 1.9 to 21.7 ± 0.4 mA/cm^2 and 0.63 ± 0.04 to 0.74 ± 0.01, respectively, as compared with the device prepared by two-step sequential deposition. A significant increase of PCE from 11.0 ± 1.5 to $15.6 \pm 0.6\%$ was achieved.[93] To simplify the deposition procedure, a low-temperature-processed CuSCN film was spin-coated onto the ITO substrate with nanocrystalline domain, followed by annealing at low temperature of 100°C. A planar inverted device of ITO/CuSCN/CH$_3$NH$_3$PbI$_3$/PC$_{61}$BM/Bis-C$_{60}$/Ag was fabricated with a compact perovskite active layer deposited by solvent engineering method. By optimizing the thickness of CuSCN of 40 nm, CuSCN-based PSCs achieved a promising efficiency of 16% and presented decent ambient stability. Interestingly, with further reduction of Ag thickness to be 20 nm, a semitransparent PSC was demonstrated with an impressive efficiency of >10%.[94] Zhao *et al.* reported the similar device structure of ITO/spin-coated CuSCN/CH$_3$NH$_3$PbI$_3$/PC$_{60}$BM/LiF/Ag, in which the perovskite active layer was deposited by solvent engineering method with a thickness of ~160 nm. A lower short current density of 15.76 ± 0.02 mA/cm^2 and FF of 0.632 ± 0.052 might be attributed to more abrupt perovskite crystalline interface and insufficient thickness of absorber less than 300 nm, showing a lower PCE of 10.8% as compared with the previous CuSCN-based PSCs.[95]

B. CuI as HTM

Copper iodide (CuI) is a wide band gap (~3.1 eV) p-type semiconducting ionic solid. On the basis of several unique properties such as suitable VB position, good optical transparency, higher hole mobility (0.5–2 cm^2/Vs) compared to CuSCN, and compatibility of solution-deposited process with the perovskite absorber, it has been successfully applied as HTM for PSCs. Christians *et al.* reported the first example by using CuI as an HTM in mesoscopic n-i-p PSCs with the best PCE of 6%. The device configuration was constructed as FTO/compact TiO$_2$/mesoporous TiO$_2$/CH$_3$NH$_3$PbI$_3$/CuI/Au. Solution-deposited CuI was applied by automated drop-casting method to form a 1.5-μm thick overlayer on top of perovskite film. Though CuI exhibited a higher electrical conductivity than spiro-OMeTAD, PSC device with this over-thick CuI resulted in a lower V_{OC} because of high recombination rate as determined by impedance spectroscopy.[96] Huangfu *et al.* prepared the CuI by means of convenient airbrush process for the PSC, which was consisted of FTO/compact TiO$_2$/mesoporous TiO$_2$/one-step spin-coated CH$_3$NH$_3$PbI$_{3-x}$Cl$_x$/CuI/Au. The thickness of mp-TiO$_2$ and CuI were separately tuned to advance the photovoltaic performance. Owing to the poor coverage of one-step spin-coated perovskite on the mp-TiO$_2$, a thick 640 nm of mp-TiO$_2$ was utilized to adsorb enough perovskite. Additionally, an over-thick perovskite layer was formed to prevent the shunt path between CuI and TiO$_2$, while a thick CuI was exploit to improve conductivity. The resultant device with a thickness of 24.8 ± 2.6 μm was demonstrated to show remarkable J_{SC} of 22.6 ± 0.8 mA/cm^2, however, a depressed efficiency of 5.6 ± 0.2% was delivered due to the poor FF and low V_{OC} of 640 ± 46 mV. As evidenced from the impedance spectroscopy, the low V_{OC} was attributed to the high recombination rate caused by the thick CuI layer.[97] Similarly, Sepalage *et al.* described the planar n-i-p PSCs with a structure of FTO/compact TiO$_2$/CH$_3$NH$_3$PbI$_3$/CuI/graphite. The planar perovskite layer was deposited using a gas-assisted spin coating method,[65] while the CuI layer and graphite layer were fabricated by doctor blading. Higher photovoltaic performance in V_{OC} was obtained compared to previously reported mesoscopic CuI-based devices, resulting in a higher efficiency of 7.5%. The improved voltage was caused by a reduced thickness of CuI film (~400 nm), however, it is still lower than the expected V_{OC} of ~1 V due to rapid recombination at the perovskite/CuI interface. On the other hand, the CuI-based devices displayed an apparent reduced *J-V* hysteresis under the DSC-like architecture. Under steady-state current measurement, the CuI-based devices showed a faster response in following the voltage steps. Rapid injection of holes from perovskite into the CuI layer prevents the capacitive current stored in the perovskite material and has positive contribution to diminish the *J–V* hysteresis.[98]

Chen *et al.* employed solution-processed CuI as an HTM in inverted planar p-i-n PSCs with the architecture of FTO/CuI/CH$_3$NH$_3$PbI$_3$/PCBM/Al. By using vapor-assisted solution process to fabricate high-quality perovskite light absorber,[99] the device showed a promising performance with a V_{OC} of 1.04 V, J_{SC} of 21.06 mA/cm^2, and PCE of 13.58%. Benefiting from the high transmittance, nanostructure morphology, and deep VB of CuI thin film, CuI-based PSCs exhibited higher V_{OC} and J_{SC} than devices using PEDOT:PSS. Similarly, under the inverted architecture, CuI-based p-i-n PSCs showed a neglected hysteresis effect and acceptable ambient stability.[100]

C. Cu$_2$O and CuO as HTM

Copper oxides, namely cuprous oxide (Cu$_2$O) and cupric oxide (CuO), are another family of well-known p-type semiconductors, which have been applied as HTM for PSCs to minimize the energy loss because of their low-lying valence bands. The photovoltaic performances of n-i-p hetero-junction planar PSCs of FTO/TiO$_2$/CH$_3$NH$_3$PbI$_3$/HTM/Au with NiO, CuI, CuSCN, spiro-OMeTAD, and Cu$_2$O as HTM are estimated *via* a new solar cell simulation software, wxAMPS (Analysis of Microelectronic and Photonic Structure). The optimized thickness of ETM and HTM are 135–145 nm and 350–450 nm, respectively, and that of perovskite absorber layer is 350–450 nm. The PCE values with various HTM display a ranking of Cu$_2$O-, CuSCN-, spiro-OMeTAD-, CuI-, and NiO-based PSC, among which the simulation of Cu$_2$O-based PSC exhibit the highest efficiency exceeding 24%.[101] The result indicates that it is possible to further increase the performance of PSCs by using copper oxide as HTM. Nejand *et al.* employed reactive magnetron sputtering process with angular rotational substrate to prepare uniform, dense, and pinhole-free crystalline layer of Cu$_2$O on the perovskite surface to facilitate extraction of holes from the perovskite layer. During the sputtering process, a 45° tilting angle of substrate against the sputtering target was conducted to create a compact and uniform Cu$_2$O layer covering on perovskite surface. In addition, the direct current (DC) power and deposition period were well controlled to prevent bombardment damage on perovskite, which may cause degradation of the perovskite to PbI$_2$. A planar n-i-p heterojunction configuration of FTO/compact TiO$_2$/CH$_3$NH$_3$PbI$_{3-x}$Cl$_x$/Cu$_2$O/Au was fabricated with a PCE of 8.93%, in which a pinhole-free perovskite light absorber was formed by the modified vapor deposition. After optimizing the thickness of the sputter-deposited Cu$_2$O layer, higher hole mobility as well as lower resistivity of Cu$_2$O film was obtained, leading to an increased PCE over 9%. The as-fabricated PSCs showed minor hysteresis due to the balance between electron and hole transport. Long-term durability of 1 month was realized because Cu$_2$O effectively hindered the water and oxygen penetration.[102]

A planar inverted p-i-n heterojunction architecture of ITO/HTM/CH$_3$NH$_3$PbI$_3$/PC$_{61}$BM/Ca/Al was experimentally demonstrated by inserting Cu_2O or CuO as HTM. Cu_2O film was fabricated by converting the CuI-spin-coated substrate *via* aqueous NaOH solution while CuO film was formed by heating Cu_2O film in air. With the addition of NH_4Cl dissolved in $CH_3NH_3PbI_3$ precursor, uniform and high coverage perovskite light absorbers were deposited on the HTM. The experimental results indicate that Cu_2O-based and CuO-based PSCs respectively display enhanced V_{OC} of 1.07 and 1.06 V and higher J_{SC} of 16.52 and 15.82 mA/cm^2, leading to the best PCE of 13.35% and 12.16%, respectively. As compared to the photovoltaic performance of PEDOT:PSS-based PSC (V_{OC} of 0.95 V, J_{SC} of 14.82 mA/cm^2, and PCE of 11.04%), enhanced V_{OC} resulted from the better matched VB of Cu_2O and CuO with that of the perovskite. Highly crystalline $CH_3NH_3PbI_3$ on Cu_2O and CuO improved the carrier transport, which increased the J_{SC}.[15a] Chatterjee *et al.* prepared a high degree of crystallinity of Cu_2O as an HTM through a successive ionic layer adsorption and reaction (SILAR) method. A similar device with the structure of ITO/Cu_2O/one-step deposited $CH_3NH_3PbI_3$/PCBM/ Al was demonstrated with an efficiency of 8.23%. Compared to the devices employing NiO or Cu:NiO HTM, Cu_2O-based PSCs exhibited a superior performance by virtue of higher mobility of Cu_2O, low energy loss conducted by scanning tunneling spectroscopy, low Cu_2O/perovskite interface recombination loss, and better perovskite crystallinity on Cu_2O film.[103] Yu *et al.* reported the inverted PSCs using ultrathin Cu_2O as p-type contact *via* thermal oxidation of sputtered Cu film. The planar device was composed of ITO/Cu_2O/sequentially deposited $CH_3NH_3PbI_3$/ PCBM/Ag, in which the thickness of Cu_2O was precisely controlled to optimize the optical transmission as well as the series resistance. As a result, the as-fabricated device employing 5 nm Cu_2O achieved the best performance with a PCE of 11.0%.[104] Sun *et al.* used facile solution-processed method to simplify the film formation of CuO_x and introduced it as an HTM into a planar inverted configuration. The solution-processed CuO_x deposited on ITO with UVO treatment exhibits a smooth surface and high transparency in the visible region. More efficient hole transport from the perovskite to CuO_x layer and lower contact resistance at the perovskite/CuO_x interface were evidenced from the time-resolved photoluminescence spectra and electrochemical impedance spectra, respectively, compared to the perovskite/PEDOT:PSS interface. The CuO_x-based device composed of $ITO/CuO_x/CH_3NH_3PbI_3/C_{60}/BCP/Ag$ showed a remarkable efficiency of 17.1% with a high photocurrent J_{SC} of 23.2 mA/cm^2, negligible hysteresis, and a superior stability compared to the PEDOT:PSS counterpart.[105] Moreover, Wang *et al.* conducted a theoretical calculation on the Cu_2O/perovskite heterojunction solar cells by using wxAMPS software. Models of inverted planar PSCs were performed with the configuration of $FTO/HTM/CH_3NH_3PbI_{3-x}Cl_x/PCBM/Al$, in which NiO_x and Cu_2O were applied as HTM, and the defect states are considered to take part in the interface recombination. An efficiency of 9.88% was first simulated for NiO_x-based device to verify the experimental results with a thickness of 500 nm for perovskite absorber layer.[62] For high-mobility Cu_2O with 10–50 nm thickness, PSCs using Cu_2O with 500-nm thick perovskite absorber was further simulated to achieve a PCE of above 13%.[106]

D. CuInS$_2$ as HTM

Copper indium disulfide ($CuInS_2$) is a promising light absorber as well as a p-type semiconductor due to its direct band gap of 1.5 eV close to the best band gap (1.45 eV) for solar cells. The valence band edge of $CuInS_2$ matches well with that of $CH_3NH_3PbI_3$ perovskite. Chen *et al.* firstly employed low-temperature-processed (250°C) $CuInS_2$ nanocrystals with a polycrystalline *via* molecular-based precursor solution as HTM for inverted device configuration. An ETM-free structure of $ITO/CuInS_2/mp$-$Al_2O_3/CH_3NH_3PbI_3/Ag$ was fabricated to exhibit an efficiency of 5.30% by optimizing the spin-coated $CuInS_2$ for two times. It was worthy noted that the IPCE response from 820 to 1000 nm benefited the photocurrent, which originated from the light harvesting from $CuInS_2$ layer; therefore, inserted $CuInS_2$ can further improve the device performance.[107] Lv *et al.* used $CuInS_2$ quantum dots as HTM in FTO/cp-TiO_2/mp-TiO_2/sequential deposited $CH_3NH_3PbI_3/CuInS_2/Au$ mesoscopic device, showing a PCE of 6.57%. When the $CuInS_2$ quantum dots with a ZnS shell layer were synthesized by cation exchange to passive their surface defects, apparent improvement in efficiency was made due to the promotion of the charge collection from perovskite to HTM. The same mesoscopic device employing $CuInS_2$-ZnS core-shell quantum dots reached an increased efficiency of 8.38%. However, the ultrathin HTM incompletely covered the rough perovskite surface made by sequential deposition, reaching an inferior performance compared to the spiro-OMeTAD-based device.[108]

E. CuAlO$_2$ as HTM

As mentioned earlier, the most widely used HTM of PEDOT:PSS in inverted structure has acidity to erode the underlying ITO, leading to a poorly stable device. A interfacial layer of CuAlO$_2$ was inserted between acidic PEDOT:PSS and ITO to prevent a direct contact. The low-temperature magnetron-sputtered amorphous CuAlO$_2$ (a:CuAlO$_2$) film exhibited ambient stability as well as good conductivity. It can serve as a buffer layer and efficient hole extraction layer. With the modification on the thickness (15 nm) and work function (5.2 eV) of a:CuAlO$_2$, the planar heterojunction device of ITO/a:CuAlO$_2$/PEDTO:PSS/CH$_3$NH$_3$PbI$_{3-x}$Cl$_x$/ PCBM/Ag using a:CuAlO$_2$/PEDOT:PSS bilayer hole conductor was fabricated to present a decent performance of a V_{OC} of 0.88 V, J_{SC} of 21.98 mA/cm^2, FF of 0.75, and PCE of 14.52%. Consequently, the introduction of the a:CuAlO$_2$ buffer layer in the PSCs showed improved stability as compared to the PEDOT:PSS-based device.[109]

F. CuS as HTM

A facile, low-temperature and low-cost solution-processed copper sulfide (CuS) was firstly demonstrated by Rao *et al.* to fabricate an efficient hole extraction layer for the planar inverted p-i-n heterojunction architecture of ITO/CuS/CH$_3$NH$_3$PbI$_3$/C$_{60}$/BCP/Ag. The spin-coated CuS nanoparticle on ITO substrate modified the surface work function from 4.9 to 5.1 eV, which matched well with the VB of CH$_3$NH$_3$PbI$_3$ perovskite (~5.4 eV). With the aid of fast-deposition crystallization method, a conformal and void-free perovskite was prepared to complete the device. By controlling the spin coating times of CuS nanoparticles for two times, a promising PCE of 16.2% including a V_{OC} of 1.02 V, J_{SC} of 22.3 mA/cm^2, and FF of 0.71 was reached.[110] Photovoltaic parameters of Cu-based HTMs for PSCs are summarized in Table 3.

Table 3. Summary of photovoltaic characteristics of PSC using Cu-based HTMs. Taken from Ref. 13 with permission of MDPI.

Type	Architecture	V_{OC} (V)	J_{SC} (mA/cm^2)	FF	PCE(%)	Ref.
CuSCN						
M	FTO/cp-TiO$_2$/mp-TiO$_2$/Psk/CuSCN/Au	0.63	14.5	0.53	4.85	85
M	FTO/cp-TiO$_2$/mp-TiO$_2$/Sb$_2$S$_3$/Psk/CuSCN/Au	0.57	17.23	0.52	5.12	86
M	FTO/cp-TiO$_2$/mp-TiO$_2$/Psk/CuSCN/Au	1.025	17.91	0.57	10.51	88
M	FTO/cp-TiO$_2$/mp-TiO$_2$/Psk/CuSCN/Au	1.016	19.7	0.62	12.4	89
M	FTO/cp-TiO$_2$/mp-TiO$_2$/Psk/CuSCN/Au	0.96	18.23	0.68	11.96	90
P	FTO/cp-TiO$_2$/Psk/CuSCN/Au	0.97	18.42	0.40	7.19	90
P	FTO/cp-TiO$_2$/Psk/CuSCN/Au	0.727	14.4	0.62	6.4	91
P	FTO/CuSCN/Psk/PCBM/Ag	0.677	8.8	0.63	3.8	61
P	ITO/CuSCN/Psk/C$_{60}$/BCP/Ag	0.97 ± 0.02	21.7 ± 0.4	0.742 ± 0.014	15.6 ± 0.6	93
P	ITO/CuSCN/Psk/PCBM/Bis-C$_{60}$/Ag	1.07 ± 0.01	19.6 ± 0.3	0.74 ± 0.03	15.6	94
P	ITO/CuSCN/Psk/PCBM/Bis-C$_{60}$/semitransparent Ag	1.06 ± 0.02	13.0 ± 0.4	0.73 ± 0.02	10.06	94
P	ITO/CuSCN/Psk/PC$_{60}$BM/LiF/Ag	1.06 ± 0.01	15.76 ± 0.02	0.632 ± 0.052	10.5 ± 0.16	95
CuI						
M	FTO/cp-TiO$_2$/mp-TiO$_2$/Psk/CuI/Au	0.55	17.8	0.62	6.0	96
M	FTO/cp-TiO$_2$/mp-TiO$_2$/Psk/CuI/Au	0.64 ± 0.046	22.6 ± 0.8	0.39 ± 0.04	5.6 ± 0.2	97
P	FTO/TiO$_2$/Psk/CuI/Graphite	0.78	16.7	0.57	7.5	98
P	FTO/CuI/Psk/PCBM/Al	1.04	21.06	0.62	13.58	100

(Continued)

Table 3. (Continued)

Type	Architecture	V_{OC} (V)	J_{SC} (mA/cm^2)	FF	PCE(%)	Ref.
Cu$_2$O						
P	(Simulation) FTO/TiO$_2$/Psk/Cu$_2$O/Au	1.2	24.75	0.82	24.4	101
P	FTO/cp-TiO$_2$/Psk/Cu$_2$O/Au	0.96	15.8	0.59	11.5	102
P	ITO/Cu$_2$O/Psk/PCBM/Ca/Al	1.07	16.52	0.755	13.35	15a
P	ITO/Cu$_2$O/Psk/PCBM/Al	0.89	16.52	0.58	8.30	103
P	ITO/Cu$_2$O/Psk/PCBM/Ag	0.952	17.5	0.662	11.03	104
P	(Simulation) FTO/Cu$_2$O/Psk/PCBM/Al	0.91	20.22	0.74	13.6	106
CuO						
P	ITO/CuO/Psk/PCBM/Ca/Al	1.06	15.82	0.725	12.16	15a
P	ITO/CuO$_x$/Psk/C$_{60}$/BCP/Ag	0.99	23.2	0.744	17.1	105
CuInS$_2$						
M	ITO/CuInS$_2$/mp-Al$_2$O$_3$/Psk/Ag	0.76	9.92	0.70	5.30	107
M	FTO/cp-TiO$_2$/mp-TiO$_2$/Psk/CuInS$_2$-ZnS/Au	0.924	18.6	0.487	8.38	108
CuAlO$_2$						
P	ITO/a:CuAlO$_2$/PEDOT:PSS/Psk/PCBM/Ag	0.88 ± 0.01	21.98 ± 0.34	0.75 ± 0.01	14.52 ± 0.58	109
CuS						
P	ITO/CuS/Psk/C$_{60}$/BCP/Ag	1.02	22.3	0.712	16.2	110

Type, "M" or "P" indicates "mesoporous" or "planar" structure respectively. In architecture, "cp" or "mp" means "compact film" or "mesoporous film".

IV. Conclusion and Perspectives

Material selection and interface engineering of ETM (or HTM) have significant impacts on PSCs performance. Suitable energy level matching with perovskite is essential for effective charge extraction of electrons or holes from organometal trihalide perovskite absorber. The latest progress in PSCs was surveyed comprehensively using alternative inorganic p-type semiconductor as potential candidates for replacing expensive HTMs. Briefly, good HTM candidates for efficient PSCs should have the following properties: (1) suitable VB position to minimize the energy loss, (2) efficient charge transporting and blocking capability, and (3) high carrier mobility. As revealed, diverse inorganic p-type semiconductor having favorable VB, including NiO, CuSCN, CuI, Cu_2O, CuO, $CuInS_2$, $CuAlO_2$ and CuS, have been introduced to demonstrate efficient PSCs with enhanced device stability. NiO is currently the most investigated material while Cu_2O is considered as a promising alternative for highly efficient PSCs due to its high hole mobility and matching energy level alignment with $MAPbI_3$ perovskite. It shall be noted that choosing an HTM with high transparency throughout the visible and near-infrared spectrum minimizes the optical loss for OPV-like architectures. For this purpose, NiO would be a promising candidate for the OPV-like PSC due to its better transparent than Cu_2O. Despite intrinsic NiO film exhibits inferior hole mobility, doping metal ion or NiO/perovskite interface modification has been demonstrated to apparently improve the devices performance. Furthermore, proper control of surface energy will affect the crystallization kinetics and film quality of the perovskite film deposited on the top of various HTMs. Combined with versatile fabrication processes such as solvent engineering, fast deposition crystallization, vapor deposition, and additives mixture, high quality of perovskite thin film can be realized. With appropriate choice of materials and fabrication process, the OPV-like device performance is apparently approaching the performance of conventional DSC-like PSC devices. The balance between charge separation, transportation, and recombination shall be considered for optimizing the internal surface area and mesoscopic p-type oxide film thickness, leading to a hysteresis-free solar cell.

For future improvement of device performance, several issues deserve further investigations. For HTM in PSCs, increasing the mobility and conductivity in p-type oxide layer is important for receiving high-efficiency PSCs. On the other hand, doping and optical absorption shall be carefully managed for achieving high overall performance. The use of metal oxides as both HTM and ETM in PSCs is an appealing future vision considering their robust behavior, long-term durability, low cost, and environmental friendly characteristics. Except for the enhancement of photovoltaic performance, flexibility, and long-term stability should be further addressed in the relevant issues. Low-temperature procedure for ETM/HTM is required to offer the feasibility on the flexible substrate. MAPbI$_3$-based PSCs using solution-processed n-type/p-type metal oxide materials have shown great promises in facile fabrication and ambient stability. Other p-type materials such as spinel, Ga-, or Cr-based oxide may find their success in PSCs.

V. References

1. Yang, W. S.; Noh, J. H.; Jeon, N. J.; Kim, Y. C.; Ryu, S.; Seo, J.; Seok, S. I. *Science* **2015**, *348*, 1234–1237.
2. Park, N.-G. *Mater. Today* **2015**, *18*, 65–72.
3. Miyata, A.; Mitioglu, A.; Plochocka, P.; Portugall, O.; Wang, J. T.-W.; Stranks, S. D.; Snaith, H. J.; Nicholas, R. J. *Nat. Phys.* **2015**, *11*, 582–587.
4. (a) Stranks, S. D.; Eperon, G. E.; Grancini, G.; Menelaou, C.; Alcocer, M. J. P.; Leijtens, T.; Herz, L. M.; Petrozza, A.; Snaith, H. J. *Science* **2013**, *342*, 341–344; (b) Li, Y.; Yan, W.; Li, Y.; Wang, S.; Wang, W.; Bian, Z.; Xiao, L.; Gong, Q. *Sci. Rep.* **2015**, *5*, 14485.
5. (a) Wehrenfennig, C.; Eperon, G. E.; Johnston, M. B.; Snaith, H. J.; Herz, L. M. *Adv. Mater.* **2014**, *26*, 1584–1589; (b) Ponseca, C. S.; Savenije, T. J.; Abdellah, M.; Zheng, K.; Yartsev, A.; Pascher, T.; Harlang, T.; Chabera, P.; Pullerits, T.; Stepanov, A.; Wolf, J.-P.; Sundström, V. *J. Am. Chem. Soc.* **2014**, *136*, 5189–5192.
6. Kojima, A.; Teshima, K.; Shirai, Y.; Miyasaka, T. *J. Am. Chem. Soc.* **2009**, *131*, 6050–6051.
7. Im, J.-H.; Lee, C.-R.; Lee, J.-W.; Park, S.-W.; Park, N.-G. *Nanoscale* **2011**, *3*, 4088–4093.
8. Kim, H.-S.; Lee, C.-R.; Im, J.-H.; Lee, K.-B.; Moehl, T.; Marchioro, A.; Moon, S.-J.; Humphry-Baker, R.; Yum, J.-H.; Moser, J. E.; Grätzel, M.; Park, N.-G. *Sci. Rep.* **2012**, *2*, 591.
9. Lee, M. M.; Teuscher, J.; Miyasaka, T.; Murakami, T. N.; Snaith, H. J. *Science* **2012**, *338*, 643–647.
10. (a) Meng, L.; You, J.; Guo, T.-F.; Yang, Y. *Acc. Chem. Res.* **2016**, *49*, 155–165; (b) Li, M.-H.; Shen, P.-S.; Wang, K.-C.; Guo, T.-F.; Chen, P. *J. Mater. Chem. A* **2015**,

3, 9011–9019; (c) Liu, T.; Chen, K.; Hu, Q.; Zhu, R.; Gong, Q. *Adv. Energy Mater.* **2016**, *6*, 1600457; (d) Yan, W.; Ye, S.; Li, Y.; Sun, W.; Rao, H.; Liu, Z.; Bian, Z.; Huang, C. *Adv. Energy Mater.* **2016**, *6*, 1600474.

11. Kim, H.-S.; Lee, C.-R.; Im, J.-H.; Lee, K.-B.; Moehl, T.; Marchioro, A.; Moon, S.-J.; Humphry-Baker, R.; Yum, J.-H.; Moser, J. E.; Grätzel, M.; Park, N.-G. *Scientific Reports* **2012**, *2*, 591.

12. (a) Yu, Z.; Sun, L. *Adv. Energy Mater.* **2015**, *5*, 1500213; (b) Salim, T.; Sun, S.; Abe, Y.; Krishna, A.; Grimsdale, A. C.; Lam, Y. M. *J. Mater. Chem. A* **2015**, *3*, 8943–8969; (c) Ameen, S.; Rub, M. A.; Kosa, S. A.; Alamry, K. A.; Akhtar, M. S.; Shin, H.-S.; Seo, H.-K.; Asiri, A. M.; Nazeeruddin, M. K. *ChemSusChem.* **2016**, *9*, 10–27; (d) Luo, S.; Daoud, W. A. *J. Mater. Chem. A* **2015**, *3*, 8992–9010; (e) Shi, S.; Li, Y.; Li, X.; Wang, H. *Mater. Horiz.* **2015**, *2*, 378–405; (f) Tong, X.; Lin, F.; Wu, J.; Wang, Z. M. *Adv. Sci.* **2016**, *3*, 1500201; (g) Swetha, T.; Singh, S. P. *J. Mater. Chem. A* **2015**, *3*, 18329–18344.

13. Li, M.-H.; Yum, J.-H.; Moon, S.-J.; Chen, P. *Energies* **2016**, *9*, 331.

14. (a) Hua, Y.; Xu, B.; Liu, P.; Chen, H.; Tian, H.; Cheng, M.; Kloo, L.; Sun, L. *Chem. Sci.* **2016**, *7*, 2633–2638; (b) Javier Ramos, F.; Ince, M.; Urbani, M.; Abate, A.; Gratzel, M.; Ahmad, S.; Torres, T.; Nazeeruddin, M. K. *Dalton Trans.* **2015**, *44*, 10847–10851; (c) Kumar, C. V.; Sfyri, G.; Raptis, D.; Stathatos, E.; Lianos, P. Perovskite *RSC Adv.* **2015**, *5*, 3786–3791.

15. (a) Zuo, C.; Ding, L. *Small* **2015**, *11*, 5528–5532; (b) Xu, B.; Bi, D.; Hua, Y.; Liu, P.; Cheng, M.; Gratzel, M.; Kloo, L.; Hagfeldt, A.; Sun, L. *Energy Environ. Sci.* **2016**, *9*, 873–877; (c) Bi, D.; Xu, B.; Gao, P.; Sun, L.; Grätzel, M.; Hagfeldt, A. *Nano Energy* **2016**, *23*, 138–144.

16. (a) Nguyen, W. H.; Bailie, C. D.; Unger, E. L.; McGehee, M. D. *J. Am. Chem. Soc.* **2014**, *136*, 10996–11001; (b) Murray, A. T.; Frost, J. M.; Hendon, C. H.; Molloy, C. D.; Carbery, D. R.; Walsh, A. *Chem. Commun.* **2015**, *51*, 8935–8938; (c) Jeon, N. J.; Lee, H. G.; Kim, Y. C.; Seo, J.; Noh, J. H.; Lee, J.; Seok, S. I. *J. Am. Chem. Soc.* **2014**, *136*, 7837–7840.

17. Lin, Y.-D.; Ke, B.-Y.; Lee, K.-M.; Chang, S. H.; Wang, K.-H.; Huang, S.-H.; Wu, C.-G.; Chou, P.-T.; Jhulki, S.; Moorthy, J. N.; Chang, Y. J.; Liau, K.-L.; Chung, H.-C.; Liu, C.-Y.; Sun, S.-S.; Chow, T. J. *ChemSusChem.* **2016**, *9*, 274–279.

18. (a) Lv, S.; Han, L.; Xiao, J.; Zhu, L.; Shi, J.; Wei, H.; Xu, Y.; Dong, J.; Xu, X.; Li, D.; Wang, S.; Luo, Y.; Meng, Q.; Li, X. *Chem. Commun.* **2014**, *50*, 6931–6934; (b) Krishna, A.; Sabba, D.; Yin, J.; Bruno, A.; Boix, P. P.; Gao, Y.; Dewi, H. A.; Gurzadyan, G. G.; Soci, C.; Mhaisalkar, S. G.; Grimsdale, A. C. *Chem. Eur. J.* **2015**, *21*, 15113–15117; (c) Abate, A.; Planells, M.; Hollman, D. J.; Barthi, V.; Chand, S.; Snaith, H. J.; Robertson, N. *Phys. Chem. Chem. Phys.* **2015**, *17*, 2335–2338; (d) Ma, Y.; Chung, Y.-H.; Zheng, L.; Zhang, D.; Yu, X.; Xiao, L.; Chen, Z.; Wang, S.; Qu, B.; Gong, Q.; Zou, D. *ACS Appl. Mat. Interfaces* **2015**, *7*, 6406–6411; (e) Wang, J.; Wang, S.; Li, X.; Zhu, L.; Meng, Q.; Xiao, Y.; Li, D. *Chem. Commun.* **2014**, *50*, 5829–5832; (f) Lv, S.; Song, Y.; Xiao, J.; Zhu, L.; Shi, J.; Wei, H.; Xu, Y.; Dong, J.; Xu, X.; Wang, S.; Xiao, Y.; Luo, Y.; Li, D.; Li, X.; Meng, Q. *Electrochim. Acta* **2015**, *182*, 733–741.

19. (a) Sung, S. D.; Kang, M. S.; Choi, I. T.; Kim, H. M.; Kim, H.; Hong, M.; Kim, H. K.; Lee, W. I. *Chem. Commun.* **2014**, *50*, 14161–14163; (b) Wang, H.; Sheikh, A. D.; Feng, Q.; Li, F.; Chen, Y.; Yu, W.; Alarousu, E.; Ma, C.; Haque, M. A.; Shi, D.; Wang, Z.-S.; Mohammed, O. F.; Bakr, O. M.; Wu, T. *ACS Photonics* **2015**, *2*, 849–855; (c) Kang, M. S.; Sung, S. D.; Choi, I. T.; Kim, H.; Hong, M.; Kim, J.; Lee, W. I.; Kim, H. K. *ACS Appl. Mat. Interfaces* **2015**, *7*, 22213–22217.

20. Gratia, P.; Magomedov, A.; Malinauskas, T.; Daskeviciene, M.; Abate, A.; Ahmad, S.; Grätzel, M.; Getautis, V.; Nazeeruddin, M. K. *Angew. Chem. Int. Ed.* **2015**, *54*, 11409–11413.

21. (a) Lim, K.; Kang, M.-S.; Myung, Y.; Seo, J.-H.; Banerjee, P.; Marks, T. J.; Ko, J. *J. Mater. Chem. A* **2016**, *4*, 1186–1190; (b) Kim, G.-W.; Kim, J.; Lee, G.-Y.; Kang, G.; Lee, J.; Park, T., 1500471; (c) Li, H.; Fu, K.; Boix, P. P.; Wong, L. H.; Hagfeldt, A.; Grätzel, M.; Mhaisalkar, S. G.; Grimsdale, A. C. *ChemSusChem.* **2014**, *7*, 3420–3425.

22. (a) Choi, H.; Park, S.; Paek, S.; Ekanayake, P.; Nazeeruddin, M. K.; Ko, J. *J. Mater. Chem. A* **2014**, *2*, 19136–19140; (b) Choi, H.; Ko, H. M.; Ko, J. *Dyes Pigm.* **2016**, *126*, 179–185.

23. Jeon, N. J.; Lee, J.; Noh, J. H.; Nazeeruddin, M. K.; Grätzel, M.; Seok, S. I. *J. Am. Chem. Soc.* **2013**, *135*, 19087–19090.

24. Liu, J.; Wu, Y.; Qin, C.; Yang, X.; Yasuda, T.; Islam, A.; Zhang, K.; Peng, W.; Chen, W.; Han, L. *Energy Environ. Sci.* **2014**, *7*, 2963–2967.

25. Ishii, A.; Jena, A. K.; Miyasaka, T. *APL Mater.* **2014**, *2*, 091102.

26. Zhang, H.; Shi, Y.; Yan, F.; Wang, L.; Wang, K.; Xing, Y.; Dong, Q.; Ma, T. *Chem. Commun.* **2014**, *50*, 5020–5022.

27. (a) Xiao, J.; Han, L.; Zhu, L.; Lv, S.; Shi, J.; Wei, H.; Xu, Y.; Dong, J.; Xu, X.; Xiao, Y.; Li, D.; Wang, S.; Luo, Y.; Li, X.; Meng, Q. *RSC Adv.* **2014**, *4*, 32918–32923; (b) Bi, D.; Mishra, A.; Gao, P.; Franckevičius, M.; Steck, C.; Zakeeruddin, S. M.; Nazeeruddin, M. K.; Bäuerle, P.; Grätzel, M.; Hagfeldt, *ChemSusChem.* **2016**, *9*, 433–438.

28. Wang, J.; Chen, Y.; Liang, M.; Ge, G.; Zhou, R.; Sun, Z.; Xue, S. *Dyes Pigm.* **2016**, *125*, 399–406.

29. Cheng, M.; Chen, C.; Xu, B.; Hua, Y.; Zhang, F.; Kloo, L.; Sun, L. *J. Energy Chem.* **2015**, *24*, 698–706.

30. Paek, S.; Rub, M. A.; Choi, H.; Kosa, S. A.; Alamry, K. A.; Cho, J. W.; Gao, P.; Ko, J.; Asiri, A. M.; Nazeeruddin, M. K. *Nanoscale* **2016**, *8*, 6335–6340.

31. (a) Lee, D.-Y.; Na, S.-I.; Kim, S.-S. *Nanoscale* **2016**, *8*, 1513–1522; (b) Wang, Z.-K.; Li, M.; Yuan, D.-X.; Shi, X.-B.; Ma, H.; Liao, L.-S. *ACS Appl. Mat. Interfaces* **2015**, *7*, 9645–9651; (c) Chang, S. H.; Lin, K.-F.; Chiu, K. Y.; Tsai, C.-L.; Cheng, H.-M.; Yeh, S.-C.; Wu, W.-T.; Chen, W.-N.; Chen, C.-T.; Chen, S.-H.; Wu, C.-G. *Sol. Energy* **2015**, *122*, 892–899.

32. (a) Zheng, L.; Chung, Y.-H.; Ma, Y.; Zhang, L.; Xiao, L.; Chen, Z.; Wang, S.; Qu, B.; Gong, Q. *Chem. Commun.* **2014**, *50*, 11196–11199; (b) Qin, P.; Kast, H.;

Nazeeruddin, M. K.; Zakeeruddin, S. M.; Mishra, A.; Bauerle, P.; Gratzel, M. *Energy Environ. Sci.* **2014**, *7*, 2981–2985.

33. Zhu, Q.; Bao, X.; Yu, J.; Zhu, D.; Qiu, M.; Yang, R.; Dong, L. *ACS Appl. Mat. Interfaces* **2016**, *8*, 2652–2657.

34. (a) Roldan-Carmona, C.; Malinkiewicz, O.; Soriano, A.; Minguez Espallargas, G.; Garcia, A.; Reinecke, P.; Kroyer, T.; Dar, M. I.; Nazeeruddin, M. K.; Bolink, H. J. *Energy Environ. Sci.* **2014**, *7*, 994–997; (b) Malinkiewicz, O.; Yella, A.; Lee, Y. H.; Espallargas, G. M.; Graetzel, M.; Nazeeruddin, M. K.; Bolink, H. J. *Nat. Photon.* **2014**, *8*, 128–132.

35. Chiang, T.-Y.; Fan, G.-L.; Jeng, J.-Y.; Chen, K.-C.; Chen, P.; Wen, T.-C.; Guo, T.-F.; Wong, K.-T. *ACS Appl. Mat. Interfaces* **2015**, *7*, 24973–24981.

36. Kwon, Y. S.; Lim, J.; Yun, H.-J.; Kim, Y.-H.; Park, T. *Energy Environ. Sci.* **2014**, *7*, 1454–1460.

37. (a) Heo, J. H.; Im, S. H.; Noh, J. H.; Mandal, T. N.; Lim, C.-S.; Chang, J. A.; Lee, Y. H.; Kim, H.-j.; Sarkar, A.; NazeeruddinMd, K.; Gratzel, M.; Seok, S. I. *Nat. Photon.* **2013**, *7*, 486–491; (b) Jeong, H.; Lee, J. K. *ACS Appl. Mat. Interfaces* **2015**, *7*, 28459–28465.

38. Bi, C.; Wang, Q.; Shao, Y.; Yuan, Y.; Xiao, Z.; Huang, J. *Nat. Commun.* **2015**, *6*.

39. (a) Seo, J.; Noh, J. H.; Seok, S. I. Rational Strategies for Efficient Perovskite Solar Cells. *Acc. Chem. Res.* **2016**, *49*, 562–572; (b) Jo, Y.; Oh, K. S.; Kim, M.; Kim, K.-H.; Lee, H.; Lce, C.-W.; Kim, D. S. *Adv. Mater. Interfaces* **2016**, *3*, 1500768.

40. Dubey, A.; Adhikari, N.; Venkatesan, S.; Gu, S.; Khatiwada, D.; Wang, Q.; Mohammad, L.; Kumar, M.; Qiao, Q. *Sol. Energ. Mat. Sol. Cells* **2016**, *145, Part 3*, 193–199.

41. Shit, A.; Nandi, A. K. *Phys. Chem. Chem. Phys.* **2016**, *18*, 10182–10190.

42. Liu, Y.; Chen, Q.; Duan, H.-S.; Zhou, H.; Yang, Y.; Chen, H.; Luo, S.; Song, T.-B.; Dou, L.; Hong, Z.; Yang, Y. *J. Mater. Chem. A* **2015**, *3*, 11940–11947.

43. (a) Wang, J.-Y.; Hsu, F.-C.; Huang, J.-Y.; Wang, L.; Chen, Y.-F. *ACS Appl. Mat. Interfaces* **2015**, *7*, 27676–27684; (b) Di Giacomo, F.; Razza, S.; Matteocci, F.; D'Epifanio, A.; Licoccia, S.; Brown, T. M.; Di Carlo, A. *J. Power Sources* **2014**, *251*, 152–156; (c) Zhang, M.; Lyu, M.; Yu, H.; Yun, J.-H.; Wang, Q.; Wang, L. *Chem. Eur. J.* **2015**, *21*, 434–439.

44. Heo, J. H.; Han, H. J.; Kim, D.; Ahn, T. K.; Im, S. H. *Energy Environ. Sci.* **2015**, *8*, 1602–1608.

45. (a) You, J.; Meng, L.; Song, T.-B.; Guo, T.-F.; Yang, Y.; Chang, W.-H.; Hong, Z.; Chen, H.; Zhou, H.; Chen, Q.; Liu, Y.; De Marco, N.; Yang, Y. *Nat. Nanotechnol.* **2016**, *11*, 75–81; (b) Chen, W.; Wu, Y.; Yue, Y.; Liu, J.; Zhang, W.; Yang, X.; Chen, H.; Bi, E.; Ashraful, I.; Grätzel, M.; Han, L. *Science* **2015**, *350*, 944–948; (c) Trifiletti, V.; Roiati, V.; Colella, S.; Giannuzzi, R.; De Marco, L.; Rizzo, A.; Manca, M.; Listorti, A.; Gigli, G. *ACS Appl. Mat. Interfaces* **2015**, *7*, 4283–4289; (d) Abdollahi Nejand, B.; Ahmadi, V.; Shahverdi, H. R. *ACS Appl. Mat. Interfaces* **2015**, *7*, 21807–21818.

46. Docampo, P.; Ball, J. M.; Darwich, M.; Eperon, G. E.; Snaith, H. J. *Nat. Commun.* **2013**, *4*, Nat. Commun. **2013**, *4*, 2761.

47. Burschka, J.; Pellet, N.; Moon, S.-J.; Humphry-Baker, R.; Gao, P.; Nazeeruddin, M.K.; Gratzel, M. *Natur* **2013**, *499*, 316–319.

48. Liu, Z.; Zhang, M.; Xu, X.; Bu, L.; Zhang, W.; Li, W.; Zhao, Z.; Wang, M.; Cheng, Y.-B.; He, H. *Dalton Trans.* **2015**, *44*, 3967–3973.

49. Liu, Z.; Zhang, M.; Xu, X.; Cai, F.; Yuan, H.; Bu, L.; Li, W.; Zhu, A.; Zhao, Z.; Wang, M.; Cheng, Y.-B.; He, H. *J. Mater. Chem. A* **2015**, *3*, 24121–24127.

50. Xu, X.; Liu, Z.; Zuo, Z.; Zhang, M.; Zhao, Z.; Shen, Y.; Zhou, H.; Chen, Q.; Yang, Y.; Wang, M. *Nano Lett.* **2015**, *15*, 2402–2408.

51. Cao, K.; Zuo, Z.; Cui, J.; Shen, Y.; Moehl, T.; Zakeeruddin, S. M.; Grätzel, M.; Wang, M. *Nano Energy* **2015**, *17*, 171–179.

52. Tian, H.; Xu, B.; Chen, H.; Johansson, E. M. J.; Boschloo, G. *ChemSusChem.* **2014**, *7*, 2150–2153.

53. Wang, K.-C.; Jeng, J.-Y.; Shen, P.-S.; Chang, Y.-C.; Diau, E. W.-G.; Tsai, C.-H.; Chao, T.-Y.; Hsu, H.-C.; Lin, P.-Y.; Chen, P.; Guo, T.-F.; Wen, T.-C. *Sci. Rep.* **2014**, *4*, 4756.

54. Jeng, J.-Y.; Chen, K.-C.; Chiang, T.-Y.; Lin, P.-Y.; Tsai, T.-D.; Chang, Y.-C.; Guo, T.-F.; Chen, P.; Wen, T.-C.; Hsu, Y.-J. *Adv. Mater.* **2014**, *26*, 4107–4113.

55. Zhu, Z.; Bai, Y.; Zhang, T.; Liu, Z.; Long, X.; Wei, Z.; Wang, Z.; Zhang, L.; Wang, J.; Yan, F.; Yang, S. *Angew. Chem. Int. Ed.* **2014**, *53*, 12571–12575.

56. Wang, K.-C.; Shen, P.-S.; Li, M.-H.; Chen, S.; Lin, M.-W.; Chen, P.; Guo, T.-F. *ACS Appl. Mat. Interfaces* **2014**, *6*, 11851–11858.

57. Chen, W.; Wu, Y.; Liu, J.; Qin, C.; Yang, X.; Islam, A.; Cheng, Y.-B.; Han, L. *Energy Environ. Sci.* **2015**, *8*, 629–640.

58. Wei-Chih, L.; Kun-Wei, L.; Tzung-Fang, G.; Jung, L. *IEEE Trans. Electron Dev.* **2015**, *62*, 1590–1595.

59. Lai, W.-C.; Lin, K.-W.; Wang, Y.-T.; Chiang, T.-Y.; Chen, P.; Guo, T.-F. *Adv. Mater.* **2016**, *28*, 3290–3297.

60. Liu, M.; Johnston, M. B.; Snaith, H. J. *Natur* **2013**, *501*, 395–398.

61. Subbiah, A. S.; Halder, A.; Ghosh, S.; Mahuli, N.; Hodes, G.; Sarkar, S. K. *J. Phys. Chem. Lett.* **2014**, *5*, 1748–1753.

62. Hu, L.; Peng, J.; Wang, W.; Xia, Z.; Yuan, J.; Lu, J.; Huang, X.; Ma, W.; Song, H.; Chen, W.; Cheng, Y.-B.; Tang, J. *ACS Photonics* **2014**, *1*, 547–553.

63. Cui, J.; Meng, F.; Zhang, H.; Cao, K.; Yuan, H.; Cheng, Y.; Huang, F.; Wang, M. *ACS Appl. Mat. Interfaces* **2014**, *6*, 22862–22870.

64. Li, E.; Guo, Y.; Liu, T.; Hu, W.; Wang, N.; He, H.; Lin, H. *RSC Adv.* **2016**, *6*, 30978–30985.

65. Huang, F.; Dkhissi, Y.; Huang, W.; Xiao, M.; Benesperi, I.; Rubanov, S.; Zhu, Y.; Lin, X.; Jiang, L.; Zhou, Y.; Gray-Weale, A.; Etheridge, J.; McNeill, C. R.; Caruso, R. A.; Bach, U.; Spiccia, L.; Cheng, Y.-B. *Nano Energy* **2014**, *10*, 10–18.

66. Xiao, M.; Gao, M.; Huang, F.; Pascoe, A. R.; Qin, T.; Cheng, Y.-B.; Bach, U.; Spiccia, L. *ChemNanoMat* **2016**, *2*, 182–188.

67. Park, I. J.; Park, M. A.; Kim, D. H.; Park, G. D.; Kim, B. J.; Son, H. J.; Ko, M. J.; Lee, D.-K.; Park, T.; Shin, H.; Park, N.-G.; Jung, H. S.; Kim, J. Y. *J. Phys. Chem. C* **2015**, *119*, 27285–27290.

68. Xiao, M.; Huang, F.; Huang, W.; Dkhissi, Y.; Zhu, Y.; Etheridge, J.; Gray-Weale, A.; Bach, U.; Cheng, Y.-B.; Spiccia, L. *Angew. Chem. Int. Ed.* **2014**, *53*, 9898–9903.

69. Etgar, L. *MRS Bull.* **2015**, *40*, 674–680.

70. Li, Y.; Ye, S.; Sun, W.; Yan, W.; Li, Y.; Bian, Z.; Liu, Z.; Wang, S.; Huang, C. *J. Mater. Chem. A* **2015**, *3*, 18389–18394.

71. (a) Unger, E. L.; Hoke, E. T.; Bailie, C. D.; Nguyen, W. H.; Bowring, A. R.; Heumuller, T.; Christoforo, M. G.; McGehee, M. D. *Energy Environ. Sci.* **2014**, *7*, 3690–3698; (b) Tress, W.; Marinova, N.; Moehl, T.; Zakeeruddin, S. M.; Nazeeruddin, M. K.; Gratzel, M. *Energy Environ. Sci.* **2015**, *8*, 995–1004.

72. Kim, H.-S.; Jang, I.-H.; Ahn, N.; Choi, M.; Guerrero, A.; Bisquert, J.; Park, N.-G. *J. Phys. Chem. Lett.* **2015**, *6*, 4633–4639.

73. Kim, J. H.; Liang, P.-W.; Williams, S. T.; Cho, N.; Chueh, C.-C.; Glaz, M. S.; Ginger, D. S.; Jen, A. K. Y. *Adv. Mater.* **2015**, *27*, 695–701.

74. Jung, J. W.; Chueh, C.-C.; Jen, A. K. Y. *Adv. Mater.* **2015**, *27*, 7874–7880.

75. Kim, J.; Lee, H. R.; Kim, H. P.; Lin, T.; Kanwat, A.; Mohd Yusoff, A. R. b.; Jang, J. *Nanoscale* **2016**, *8*, 9284–9292.

76. Park, J. H.; Seo, J.; Park, S.; Shin, S. S.; Kim, Y. C.; Jeon, N. J.; Shin, H.-W.; Ahn, T. K.; Noh, J. H.; Yoon, S. C.; Hwang, C. S.; Seok, S. I. *Adv. Mater.* **2015**, *27*, 4013–4019.

77. Seo, S.; Park, I. J.; Kim, M.; Lee, S.; Bae, C.; Jung, H. S.; Park, N.-G.; Kim, J. Y.; Shin, H. *Nanoscale* **2016**, *8*, 11403–11412.

78. Yin, X.; Que, M.; Xing, Y.; Que, W. *J. Mater. Chem. A* **2015**, *3*, 24495–24503.

79. Yin, X.; Chen, P.; Que, M.; Xing, Y.; Que, W.; Niu, C.; Shao, J. *ACS Nano* **2016**, *10*, 3630–3636.

80. Zhang, H.; Cheng, J.; Lin, F.; He, H.; Mao, J.; Wong, K. S.; Jen, A. K. Y.; Choy, W. C. H. *ACS Nano* **2015**, *10*, 1503–1511.

81. Hsu, H.-Y.; Wang, C.-Y.; Fathi, A.; Shiu, J.-W.; Chung, C.-C.; Shen, P.-S.; Guo, T.-F.; Chen, P.; Lee, Y.-P.; Diau, E. W.-G. *Angew. Chem. Int. Ed.* **2014**, *53*, 9339–9342.

82. Corani, A.; Li, M.-H.; Shen, P.-S.; Chen, P.; Guo, T.-F.; El Nahhas, A.; Zheng, K.; Yartsev, A.; Sundström, V.; Ponseca, C. S. *J. Phys. Chem. Lett.* **2016**, *7*, 1096–1101.

83. Lin, M.-W.; Wang, K.-C.; Wang, J.-H.; Li, M.-H.; Lai, Y.-L.; Ohigashi, T.; Kosugi, N.; Chen, P.; Wei, D.-H.; Guo, T.-F.; Hsu, Y.-J. *Adv. Mater. Interfaces* **2016**, *3*, 1600135.

84. Nilushi, W.; Thomas, D. A. *Semicond. Sci. Technol.* **2015**, *30*, 104002.

85. Ito, S.; Tanaka, S.; Vahlman, H.; Nishino, H.; Manabe, K.; Lund, P. *ChemPhysChem* **2014**, *15*, 1194–1200.

86. Ito, S.; Tanaka, S.; Manabe, K.; Nishino, H. *J. Phys. Chem. C* **2014**, *118*, 16995–17000.

87. Murugadoss, G.; Mizuta, G.; Tanaka, S.; Nishino, H.; Umeyama, T.; Imahori, H.; Ito, S. *APL Mater.* **2014**, *2*, 081511.

88. Ito, S.; Tanaka, S.; Nishino, H. *Chem. Lett.* **2015**, *44*, 849–851.

89. Qin, P.; Tanaka, S.; Ito, S.; Tetreault, N.; Manabe, K.; Nishino, H.; Nazeeruddin, M. K.; Grätzel, M. *Nat. Commun.* **2014**, *5*, 3834.

90. Ito, S.; Tanaka, S.; Nishino, H. *J. Phys. Chem. Lett.* **2015**, *6*, 881–886.

91. Chavhan, S.; Miguel, O.; Grande, H.-J.; Gonzalez-Pedro, V.; Sanchez, R. S.; Barea, E. M.; Mora-Sero, I.; Tena-Zaera, R. *J. Mater. Chem. A* **2014**, *2*, 12754–12760.

92. Ma, Y.; Zheng, L.; Chung, Y.-H.; Chu, S.; Xiao, L.; Chen, Z.; Wang, S.; Qu, B.; Gong, Q.; Wu, Z.; Hou, X. *Chem. Commun.* **2014**, *50*, 12458–12461.

93. Ye, S.; Sun, W.; Li, Y.; Yan, W.; Peng, H.; Bian, Z.; Liu, Z.; Huang, C. *Nano Lett.* **2015**, *15*, 3723–3728.

94. Jung, J. W.; Chueh, C.-C.; Jen, A. K. Y. *Adv. Energy Mater.* **2015**, *5*, 1500486.

95. Zhao, K.; Munir, R.; Yan, B.; Yang, Y.; Kim, T.; Amassian, A. *J. Mater. Chem. A* **2015**, *3*, 20554–20559.

96. Christians, J. A.; Fung, R. C. M.; Kamat, P. V. *J. Am. Chem. Soc.* **2014**, *136*, 758–764.

97. Huangfu, M.; Shen, Y.; Zhu, G.; Xu, K.; Cao, M.; Gu, F.; Wang, L. *Appl. Surf. Sci.* **2015**, *357, Part B*, 2234–2240.

98. Sepalage, G. A.; Meyer, S.; Pascoe, A.; Scully, A. D.; Huang, F.; Bach, U.; Cheng, Y.-B.; Spiccia, L. *Adv. Funct. Mater.* **2015**, *25*, 5650–5661.

99. Chen, Q.; Zhou, H.; Hong, Z.; Luo, S.; Duan, H.-S.; Wang, H.-H.; Liu, Y.; Li, G.; Yang, Y. *J. Am. Chem. Soc.* **2014**, *136*, 622–625.

100. Chen, W.-Y.; Deng, L.-L.; Dai, S.-M.; Wang, X.; Tian, C.-B.; Zhan, X.-X.; Xie, S.-Y.; Huang, R.-B.; Zheng, L.-S. *J. Mater. Chem. A* **2015**, *3*, 19353–19359.

101. Hossain, M. I.; Alharbi, F. H.; Tabet, N. *Sol. Energy* **2015**, *120*, 370–380.

102. Nejand, B. A.; Ahmadi, V.; Gharibzadeh, S.; Shahverdi, H. R. *ChemSusChem* **2016**, *9*, 302–313.

103. Chatterjee, S.; Pal, A. J. *J. Phys. Chem. C* **2016**, *120*, 1428–1437.

104. Yu, W.; Li, F.; Wang, H.; Alarousu, E.; Chen, Y.; Lin, B.; Wang, L.; Hedhili, M. N.; Li, Y.; Wu, K.; Wang, X.; Mohammed, O. F.; Wu, T. *Nanoscale* **2016**, *8*, 6173–6179.

105. Sun, W.; Li, Y.; Ye, S.; Rao, H.; Yan, W.; Peng, H.; Li, Y.; Liu, Z.; Wang, S.; Chen, Z.; Xiao, L.; Bian, Z.; Huang, C. *Nanoscale* **2016**, *8*, 10806–10813.

106. Yan, W.; Zhonggao, X.; Jun, L.; Xinwei, W.; Yiming, L.; Chuan, L.; Shengdong, Z.; Hang, Z. *Semicond. Sci. Technol.* **2015**, *30*, 054004.

107. Chen, C.; Li, C.; Li, F.; Wu, F.; Tan, F.; Zhai, Y.; Zhang, W. *Nanoscale Res. Lett.* **2014**, *9*, 1–8.

108. Lv, M.; Zhu, J.; Huang, Y.; Li, Y.; Shao, Z.; Xu, Y.; Dai, S. *ACS Appl. Mat. Interfaces* **2015**, *7*, 17482–17488.

109. Igbari, F.; Li, M.; Hu, Y.; Wang, Z.-K.; Liao, L.-S. *J. Mater. Chem. A* **2016**, *4*, 1326–1335.

110. Rao, H.; Sun, W.; Ye, S.; Yan, W.; Li, Y.; Peng, H.; Liu, Z.; Bian, Z.; Huang, C. *ACS Appl. Mat. Interfaces* **2016**, *8*, 7800–7805.

5 Hole Conductor Free Organometal Halide Perovskite Solar Cells: Properties and Different Architectures

Sigalit Aharon and Lioz Etgar*

The Hebrew University of Jerusalem, Institute of Chemistry,
Casali Center for Applied Chemistry, Jerusalem, Israel
*lioz.etgar@mail.huji.ac.il

Table of Contents

List of Abbreviations

AFM	atomic force microscopy
BL	blocking layer
cAFM	conductive atomic force microscopy
CV	capacitance-voltage
DMF	N,N-dimethylformamide
FF	fill factor
HCL	hole conducting layer
IMPS	incident modulated photovoltage spectroscopy
IPCE	incident photon-to-current conversion efficiency
Jsc	short-circuit current density
MAI	methylammonium iodide
$MAPbI_3$	methylammonium lead iodide
PCE	power conversion efficiency
PSC	perovskite solar cells
PV	photovoltaic
RMS	root mean square roughness
SEM	scanning electron microscopy
SPV	surface photovoltage
TCO	transparent conductive oxide
TiDIP	diisopropoxidebis(Acetylacetonate)
V_{oc}	open-circuit voltage
XHR-SEM	extra high resolution scanning electron microscopy
XRD	X-ray diffraction
ΔCPD	the contact potential difference
τr	recombination lifetime

I. Introduction

This chapter focuses on the important property of hybrid organic–inorganic perovskites to be used simultaneously as light harvester and as a hole conductor. This property was discovered about 5 years ago,[1] and since then many papers and patents were published presenting the mechanism, different optical properties, and various solar cell configurations of hole conducting layer (HCL)-free perovskite solar cells (PSCs). In this chapter, the evolution and analysis of high-efficiency HCL-free PSCs will be presented. Deposition processes, chemical treatments, and solar cell configurations are discussed. The potential of high open-circuit voltage in a single PSC without hole conductor is demonstrated with the possibility to use various metal oxides (some of them as scaffolds) will be reviewed. Finally, a unique spray deposition technique of the perovskite layer is presented. This deposition method permits fast and simple fabrication process of the perovskite layer with thickness of micrometer. The thick perovskite layer was utilized for the fabrication of efficient HCL-free PSCs.

II. High-Efficiency Hole-Conductor-Free Perovskite Solar Cells: Optimization of Key Parameters

The preparation of HCL-free PSCs is similar to the preparation of standard PSCs but without the HCL. Yet, due to the direct contact between the metallic back contact (i.e., gold) and the perovskite, there was a need to adjust the SC's structure to the absence of HCL. Here, four key parameters that mainly influence the performance of HCL-free PSCs are discussed. (1) The blocking layer (BL) deposition, (2) PbI_2 deposition before dipping in methylammonium iodide (MAI) solution, (3) mesoporous TiO_2 film thickness, and (4) postdeposition antisolvent treatment of the perovskite layer prior to the back-contact deposition.[2,3]

The solar cell structure and its energy level diagram are shown in Figure 1a and b respectively, which consists of a dense TiO_2 BL following by the deposition of mesoporous TiO_2 film, perovskite film, and gold as the back contact.

The BL has an important role as it blocks the back-route of electrons from the transparent conductive oxide (TCO) to mesoporous TiO_2. The BL is fabricated by spin coating a solution of a diluted titanium diisoprop oxidebis(acetylacetonate) (TiDIP) in ethanol. The influences of (1) changing the BL spin velocity as well as (2) the TiDIP precursor solution concentration on the photovoltaic (PV) performance were investigated. A too

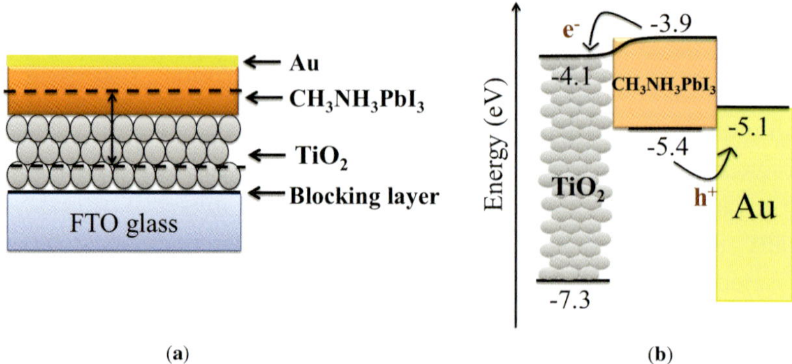

(a)   (b)

Figure 1. (a) The structure of the hole-conductor-free TiO_2/MAPbI$_3$ perovskite-based solar cells. The arrow indicates the depletion region observed at the TiO_2/MAPbI$_3$ junction. (b) Energy level diagram of the discussed solar cell which shows the charge separation process. The position of the energy levels are taken from literature.[1] Reproduced from Ref. 2 with permission of the Royal Society of Chemistry.

Table 1. Efficiencies of the hole-conductor-free PSCs at different BL conditions, i.e., spin velocity (rpm) and at various TiDIP concentrations. Taken from Ref. 2 with permission of the Royal Society of Chemistry.

Volume fraction of TiDIP (%)	Spin velocity 500 rpm (%)	Spin velocity 1500 rpm (%)	Spin velocity 2500 rpm (%)
0.05	4.5	3.0	3.0
0.10	4.2	7.0	2.0
0.15	5.0	7.2	2.2
0.20	3.0	5.4	6.5

thin BL is expected to be fragmented; thus, it cannot block the back-route of electrons. In parallel, thick BL might block the desired forward-route of electrons from TiO_2 to the TCO. Table 1 presents the PCE of cells made in different conditions. In order to study the influence of the BL conditions on the PV performance all the other parameters were held constant.

The best power conversion efficiency (PCE) was observed for spin velocity of 1500 rounds per minute (rpm) and 0.15% volume fraction of TiDIP in ethanol. Looking at the results, it was observed that for each concentration, there is an optimal spin velocity. At low TiDIP concentrations, the best PCE was observed at low spin velocity. The spin velocity determines the BL thickness, so when using low TiDIP concentration, a thick BL is preferred (i.e., low spin velocity). On the other hand, at a high concentration of TiDIP, the best efficiency was observed at the high spin velocity. The high spin velocity enables achieving a thin, uniform BL, so that a high TiDIP concentration can be used. In the case of high TiDIP concentration and low spin velocity, a thick and condensed BL was formed resulting in low PV performance.

The second factor which was investigated is related to the perovskite deposition. The deposition of the lead halide iodide perovskite on the mesoporous TiO_2 was done based on the two-step deposition method (spin and dip) described earlier[4] but with several modifications which assist in controlling the lead halide iodide perovskite deposition on the mesoporous TiO_2. The first step was spinning PbI_2 on the mesoporous TiO_2 film and annealing at 70 °C, while the second step was dipping the PbI_2 electrode into MAI solution. After the PbI_2 reacts with the MAI, the methylammonium lead iodide ($MAPbI_3$) was formed. An important factor in this deposition technique is the wait time period after dropping the PbI_2 on the mesoporous TiO_2 film, before spinning. This factor dramatically influences the filtration of the PbI_2 solution into the mesoporous TiO_2. Thus, the wait time parameter is investigated and its influence on the PV performance of these cells is observed. The results are summarized in Figure 2. The highest efficiency (8.0%) was achieved at a wait time of 3 min.

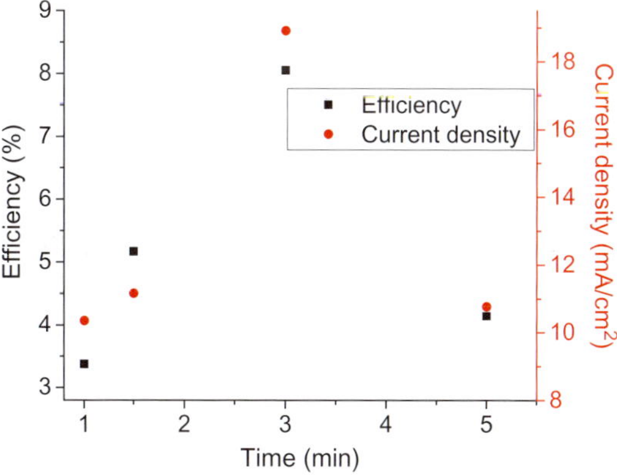

Figure 2.   The current density and the efficiency of the cells versus the wait time (minute) between dropping the PbI_2 until the beginning of the spinning. Taken from Ref. 2 with permission of the Royal Society of Chemistry.

During 3 min of waiting time, the PbI_2 creates a uniform coating on the mesoporous TiO_2 surface while a longer waiting time (5 min) causes the evaporation of the PbI_2 solvent. Solvent evaporation before spin coating could create a nonuniform coating of the PbI_2 on the mesoporous TiO_2 surface. Conversely, a too-short wait time does not allow full filtration of PbI_2 into the mesoporous TiO_2, thus might lead to internal cutoffs in the cell.

The third factor discussed here is the influence of the mesoporous TiO_2 film thickness on the solar cell performance. This factor was investigated by making hole-conductor-free perovskite-based solar cells using different thickness of mesoporous TiO_2 films while keeping the $MAPbI_3$ perovskite film thickness constant (same deposition parameters were used for the perovskite deposition). To achieve different TiO_2 film thicknesses, the TiO_2 paste was diluted by ethanol at various ratios. The various dilutions produced different viscosities of TiO_2 dispersions, which lead to different thicknesses of TiO_2 films (Figure 3a–e). The thickness of the mesoporous TiO_2 film was measured by a Dektak 150 profiler, and the results are shown in Figure 3e. Figure 3a–d presents extra high resolution scanning electron microscopy (XHR-SEM) images of the various TiO_2 thicknesses in the complete set of $TiO_2/MAPbI_3$ PSCs. It can be observed that the $MAPbI_3$ formed an over layer on top of the TiO_2 film. Probably some of the $MAPbI_3$ penetrated into the TiO_2 film; however, the thick over layer of the $MAPbI_3$ film (300 ± 50 nm) indicates that most of the perovskite is growing on top of the TiO_2 film. Table 2 summarizes the results obtained for the solar cells made with various TiO_2 thicknesses.

The highest efficiency was observed for cells made with mesoporous TiO_2 film of 620 ± 25 nm thickness. In these cells, the current density and the open-circuit voltage were higher than the other cells made with different thicknesses of TiO_2. The lowest efficiencies were observed for the thick TiO_2 film (750 ± 25 nm) and for the thin TiO_2 film (400 ± 25 nm). In addition, the fill factor (FF) and the current density for these cells were also the lowest compared with the other TiO_2 thicknesses.

To understand the reason for the difference in PV performance observed when using a variety of TiO_2 film thicknesses, capacitance–voltage (CV) measurements were performed. In the previous work,[5] it was

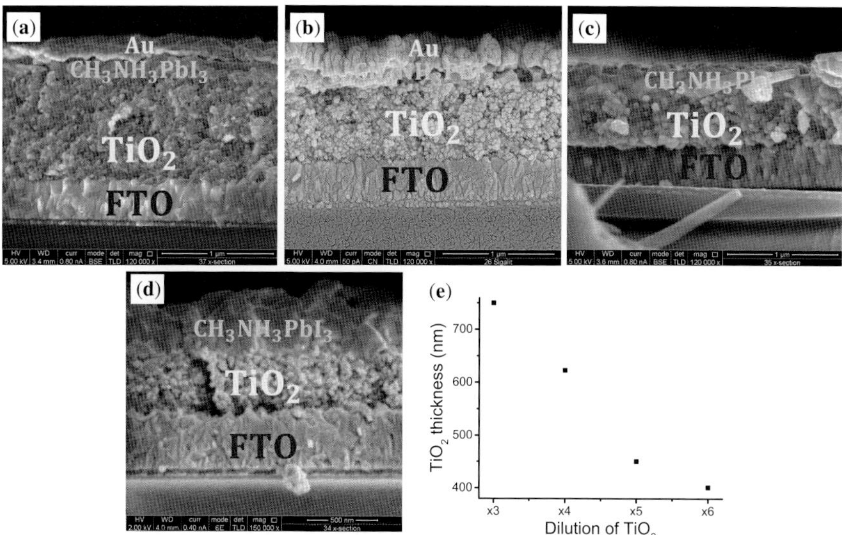

Figure 3. XHR-SEM of the different TiO_2 thickness in the complete TiO_2/$CH_3NH_3PbI_3$ perovskite-based solar cells. (**a**) The thicker TiO_2 film corresponds to 3 times dilution see Figure 3e. (**b**) 4 times dilution. (**c**) 5 times dilution. (**d**) 6 times dilution. (**e**) TiO_2 thickness (was measured by surface profiler) vs TiO_2 paste dilution factor. Reproduced from Ref. 2 with permission of the Royal Society of Chemistry.

Table 2. PV performance of the hole-conductor-free TiO_2/$MAPbI_3$ perovskite-based solar cells obtained for various TiO_2 film thicknesses. Taken from Ref. 2 with permission of the Royal Society of Chemistry.

Thickness of TiO_2 film (nm) ± 25 nm	V_{oc} (V)	I_{sc} (mA/cm^2)	FF	Efficiecny (%)
750	0.66	12.47	0.45	3.7
620	0.73	16.03	0.61	7.2
450	0.68	13.27	0.63	5.7
400	0.65	10.47	0.55	3.8

FF, fill factor.

reported that a depletion layer is created in the $MAPbI_3$/TiO_2 junction. This depletion layer assists charge separation and inhibits the back movement of electrons from the TiO_2 into the $MAPbI_3$ film. To estimate the depletion region width, Mott Schottky analysis[6] was performed on the $TiO_2/MAPbI_3$ heterojunction solar cells.

The capacitance at the junction which is described in eq 1 is calculated from the depletion approximation.[7] The depletion approximation implies that there are no free charge carriers in the space charge region at the junction under investigation.

$$\frac{1}{C^2} = \frac{2}{\varepsilon\varepsilon_0 qA^2 N}(V_{bi} - V)$$ (1)

where C is the measured capacitance, A is the active area, V is the applied bias, ε is the static permittivity, ε_0 is the permittivity of free space, q is elementary charge, and N is the doping density of the donor. The static permittivity of $MAPbI_3$ was measured and calculated to be 30.[8,9] From the slope of the Mott Schottky plot in the linear regime, the net doping density of the $MAPbI_3$ film can be calculated. Moreover, the intersection of the linear regime in the Mott Schottky plot with the x axis determines the built-in voltage, which suppresses the back reaction of electrons from the TiO_2 film toward the $MAPbI_3$ film.

The depletion width is calculated according to[10]:

$$W_{p,n} = \frac{1}{N_{a,d}}\sqrt{\frac{2\varepsilon V_{bi}}{q\left(\dfrac{1}{N_a + N_d}\right)}}$$ (2)

where N_a and N_d are the doping densities of the acceptor and the donor, respectively. Literature values for doping density in nanoporous TiO_2 start at $N_a = 1 \times 10^{16} \cdot cm^{-3}$.[11–13]

Table 3 summarizes the calculated W_n, W_p, total depletion width (W_t), and the built-in potential (V_{bi}) for the various TiO_2 film thicknesses. The depleted fraction of TiO_2, which is the ratio between the part in the TiO_2 film that is depleted and the total TiO_2 thickness.

The correlation between the PCE for the various TiO_2 film thicknesses and the depleted fraction of the TiO_2 film can be observed in Figure 4. In the case of the highest efficiency, the total depletion width (W_t) is maximal, and equals 395 nm. Moreover, half of the TiO_2 film is

Table 3. Calculated depletion width and the built-in potential for various TiO_2 film thicknesses. Taken from Ref. 2 with permission of the Royal Society of Chemistry.

Thickness of $TiO_2 \pm 25$ nm	Depleted TiO_2*	V_{oc} (V)	V_{bi} (V)	Efficiency (%)	W_t (nm)	W_n (nm)	W_p (nm)
750	0.12	0.66	0.87	3.8	223	88	135
620	0.49	0.73	0.74	7.2	395	306	89
450	0.31	0.68	0.70	5.7	239	140	99
400	0.10	0.65	0.73	3.7	194	44	150

*The depleted TiO_2 is the fraction of the part in the TiO_2 film that is depleted.

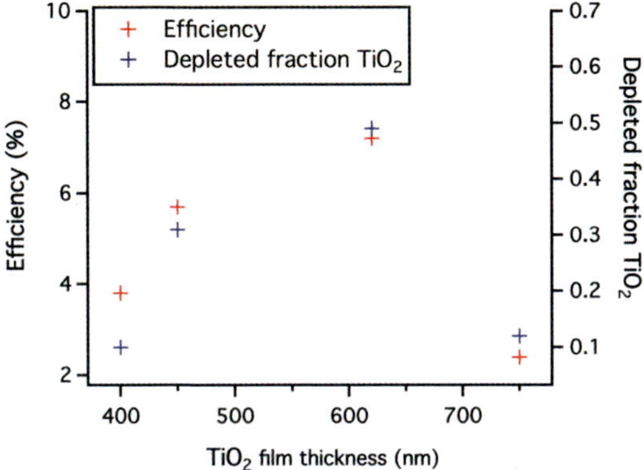

Figure 4. Efficiency and depleted fraction of the TiO_2 as a function of the TiO_2 film thickness. Taken from Ref. 2 with permission of the Royal Society of Chemistry.

depleted (depleted fraction equals to 0.49), suggesting that the depletion region indeed assists in the charge separation and suppresses the back movement of electrons, and consequently increases the PCE of the cells. For the low efficiencies, the W_t is the lowest, and the depleted fraction of the TiO_2 is around 0.1, which means that only 10% of the total TiO_2 thickness is depleted. Further, there is good agreement between the open-circuit voltage and the built-in potential observed from the Mott Schottky plot. As a conclusion, there is a very good agreement between the depletion region width and the PCE observed.

The J–V curves and the incident photon to current efficiency (IPCE) of the hole-conductor-free PSCs using various TiO_2 film thicknesses are presented in Figure 5a and b.

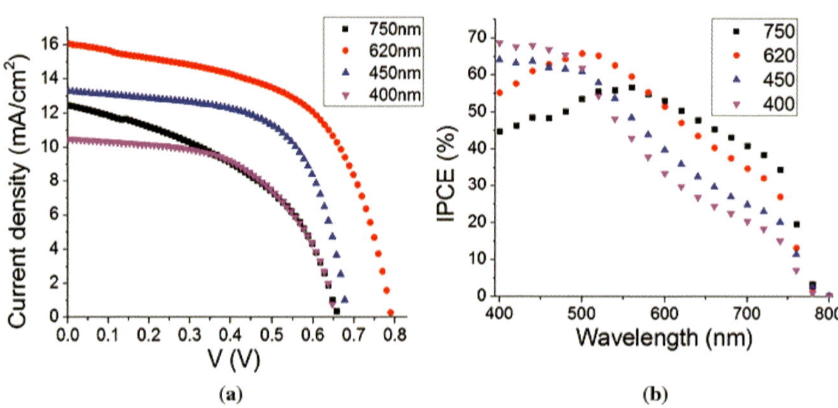

Figure 5. (**a**) *J–V* curves of the different TiO$_2$ film thickness for the hole-conductor-free PSCs and their (**b**) IPCE spectra. Taken from Ref. 2 with permission of the Royal Society of Chemistry.

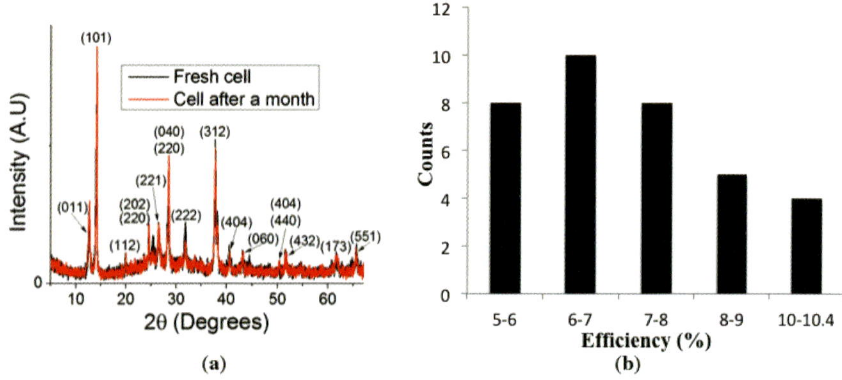

Figure 6. (**a**) XRD of fresh cell and the same cell after a month. About two third of the peaks belong to the Orthorhombic structure of the MAPbI$_3$ perovskite, while one third belongs to the tetragonal structure of the MAPbI$_3$ perovskite. (**b**) Histogram of the MAPbI$_3$/TiO$_2$ heterojunction solar cells efficiency. The average efficiency is 7.7%, with more than 11% of the cells show efficiencies above 10%. Taken from Ref. 2 with permission of the Royal Society of Chemistry.

Figure 6a presents X-ray diffraction (XRD) of a fresh cell and the same cell (which stayed in the dark) after a month. It can be seen that two-thirds of the peaks belong to the Orthorhombic MAPbI$_3$ perovskite structure, and one-third belongs to the tetragonal MAPbI$_3$ perovskite structure.[8] Moreover the peaks are completely matched between the two measurements, suggesting that there is no change in the crystallographic structure of the perovskite over time, which indicates the high stability of these hole-conductor-free PSCs. Figure 6b shows the reproducibility of the MAPbI$_3$/TiO$_2$ heterojunction solar cells made using the optimum conditions described earlier (TiO$_2$ thickness, wait time, BL condition). More than 45% of the cells show efficiencies above 8%, while more than 11% show efficiencies above 10%. The average PCE is 7.7%.

Using the optimum conditions discussed earlier, HCL-free solar cell PCE of 10.85% with V_{oc} of 0.84 V, FF of 68%, and J_{sc} of 19 mA/cm^2 is demonstrated. Its J–V curve is presented in Figure 7a. The corresponding IPCE spectra is shown in Figure 7b, it reaches its maximum of 88% at 400–550 nm wavelengths and gradually drops at longer wavelengths corresponding to its absorption spectrum. Without a hole conductor, the charge carrier at the gold back contact aren't extracted efficiently, which influence the carrier collection at longer wavelength. The integration of the IPCE spectra gives current density of 17.9 mA/cm^2 as shown in Figure 7b, there is good agreement with the current density measured by the solar simulator.

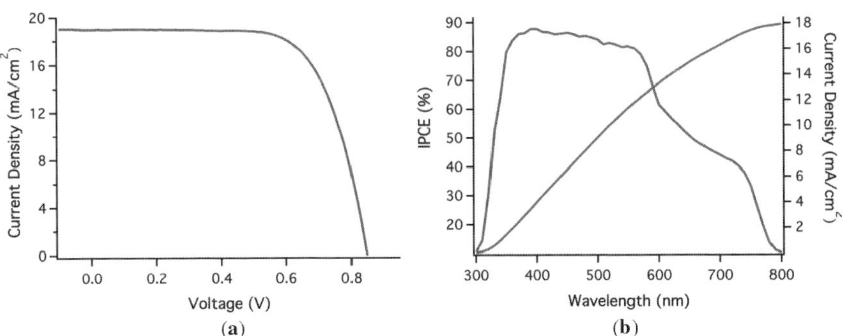

Figure 7. (a) Current density–voltage curve of the champion hole-conductor-free CH$_3$NH$_3$PbI$_3$/TiO$_2$ heterojunction solar cell; (b) The corresponding IPCE spectrum and its integrated current density. Taken from Ref. 2 with permission of the Royal Society of Chemistry.

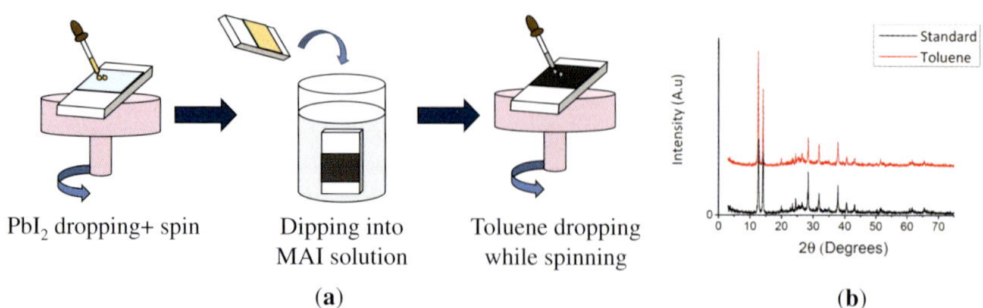

Figure 8. (**a**) Schematic presentation of the antisolvent treatment for the perovskite film deposition. (**b**) XRD spectra of the standard perovskite film and the perovskite film with the toluene treatment. (**c**) The structure of the HCL-free perovskite-based solar cell. Taken from Ref. 3 with permission of the American Chemical Society.

An important interface in the HCL-free PSCs is the perovskite/back-contact interface. In order to improve this interface, an antisolvent treatment was applied in this solar cell configuration. As shown in Figure 8, the antisolvent treatment of the perovskite film improves the film roughness and enhances its conductivity. Figure 8a presents a schematic illustration of the $MAPbI_3$ deposition process. The perovskite deposition process is based on the two-step deposition method as described earlier[4,14] with the addition of antisolvent (toluene) treatment on the perovskite film. Enhancement in the PV performance as a result of the toluene treatment is observed as discussed in the figure. Figure 8b shows the XRD spectra of perovskite film without the toluene treatment (Standard) and with the toluene treatment. No observable variations were recognized in the XRD spectra. This suggests that no crystallographic changes occurred as a result of the antisolvent treatment.

Toluene treatment of the perovskite film was previously reported as increasing the PV performance of perovskite-based solar cells with HCL.[15] Because no HCL is used in this case, the interface of the perovskite with the metal contact is significantly important and can play an important role in improving the PV performance of these cells. The JV curve of the HCL-free perovskite-based SCs is presented in Figure 9a with V_{oc} of 0.91 V, FF of 0.65, and J_{sc} of 19 mA/cm^2 corresponding to PCE of 11.2%. The normalized IPCE curve is presented at the inset of

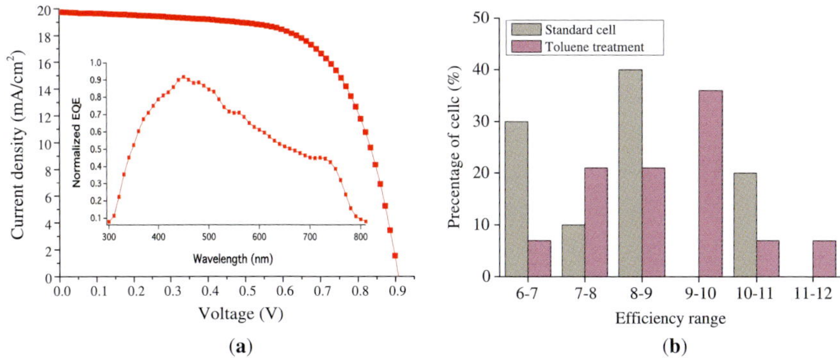

(a)   (b)

Figure 9. (a) JV curve of a toluene-treated HCL-free perovskite-based SC. Inset-IPCE curve of the corresponding cell. (b) Statistics of the efficiencies of toluene-treated and nontreated cells. The statistical analysis was performed for a total number of 28 electrodes which are equivalent to 84 cells. Taken from Ref. 3 with permission of the American Chemical Society.

Figure 9a. The integration over the IPCE spectrum gives current density of 16.2 mA/cm^2, which is in good agreement with the J_{sc} obtained by solar simulator. Figure 9b presents statistics for standard cells and for cells with toluene treatment. It is noted that the average PCE of the standard cells is $8 \pm 1\%$, while the average PCE of the toluene-treated cells is $9 \pm 1\%$. It is clear that the toluene treatment improves the PV performance of the HCL-free cells. The J_{sc} and the FF are slightly changed, while the V_{oc} is clearly affected by this treatment. The V_{oc} is larger by 0.05 V in average in the case of the toluene-treated cells. This enhancement can be related to the improved morphology and conductivity as discussed below. In order to prove the reliability of the results, a statistical analysis was performed as described by Buriak and coworkers.[16] First, the Z score is determined according to:

$$Z = \frac{X_1 - X_2}{\sigma / \sqrt{N}} \tag{3}$$

where X_2 is the average PCE of the nontreated cells, X_1 is the average PCE of the treated cells, σ is the standard deviation of the treated cells, and N is the number of cells. Using these parameters, the Z score was equal to 2.5, which correspond to p value of 0.012. Since the p value is lower than 0.05 it can be concluded with 95% confidence that the average PCE of cells fabricated using the toluene treatment is greater than the average PCE of cells without toluene treatment. This further supports the statistical reliability of the results.

So as to investigate the morphology and the electronical effects of the toluene treatment on the perovskite film, SEM, atomic force microscopy (AFM), conductive AFM (cAFM), and surface photovoltage (SPV) measurements were performed. Top-view SEM images are presented in Figure 10a and d for the toluene-treated and nontreated samples, respectively. The red circles indicate the pinholes observed in the perovskite

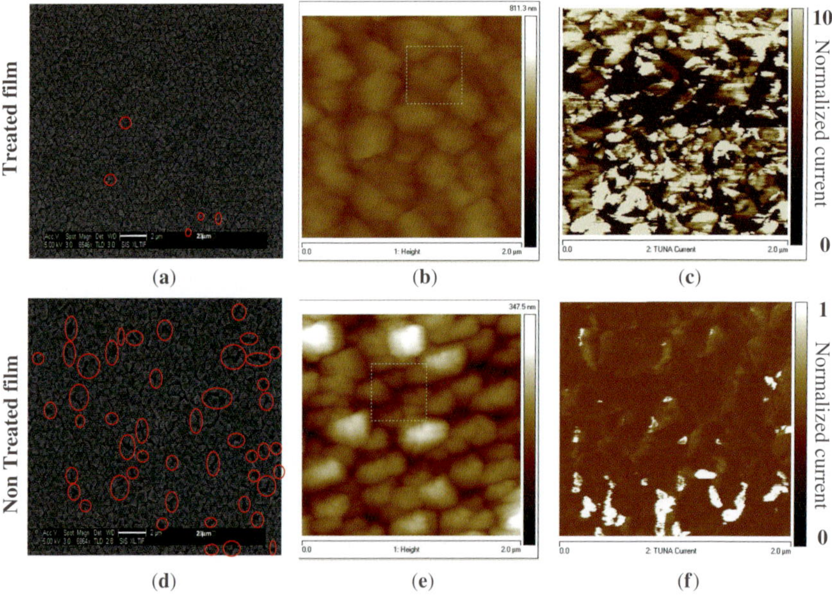

Figure 10. SEM figures of (**a**) toluene-treated solar cell and (**d**) standard solar cell. The red circles indicate pinholes in the perovskite film. (**b**) Morphology AFM images of toluene-treated and (**e**) standard solar cells. The RMS for standard cell is 40 nm and for toluene-treated cell is 30 nm. Current mapping was measured by cAFM without bias. (**c**) Conductivity of the toluene-treated perovskite film; (**f**) conductivity of the standard perovskite film. Taken from Ref. 3 with permission of the American Chemical Society.

film. It can be seen that after the toluene treatment, there are fewer pin-holes than in the nontreated perovskite film. It can be concluded that the perovskite coverage is improved as a result of this treatment. In addition, AFM measurements were performed to observe the root-mean-square (RMS) roughness of the perovskite film surface. The RMS in the case of toluene-treated film is 30 nm, while for the nontreated film, the RMS is 40 nm. This 10-nm gap indicates that in addition to the better coverage achieved by the toluene treatment, a better layer texture is achieved (the surface roughness is smaller after the perovskite is treated with toluene). Since the perovskite film is directly attached to the metal contact in the HCL-free configuration, the better coverage and the lower RMS of the toluene-treated cells contribute to better PV performance. Figure 10b and e shows AFM morphology and the corresponding cAFM measurements (current mapping) of the toluene-treated and nontreated $MAPbI_3$ perovs-kite films (Figure 10c and f). The conductivity of the perovskite film treated with toluene is larger (by a factor of 10 over the conductive grains) than the nontreated perovskite film (Figure 10c vs f). No bias was applied during the cAFM measurements. In the toluene-treated film, most of the grains are conductive (Figure 10c), while in the nontreated perovskite film the average conductivity is much lower with slightly higher conductivity at the grain boundaries (Figure 10f).

Figure 11a and b show the JV plots of a single conductive grain, in forward and reverse scans of the nontreated and toluene-treated perovskite films, respectively. The inset of Figure 11a and b presents the JV plots of a single nonconductive grain, forward and reverse scans for the two dif-ferent treatments. Several conclusions can be extracted from this measure-ment. (1) No hysteresis is observed in the toluene-treated film compared to the nontreated film. Clearly, the toluene treatment suppresses the hys-teresis. (2) In the case of the nontreated film (Figure 11a), the cAFM measurements show direct experimental observation of the memory prop-erties of the perovskite.[17] (3) Looking at the JV slope in the linear region for both films (Figure 11a and b), it can be observed that the slope in the IV plot is smaller in the toluene-treated perovskite than in the nontreated perovskite. The difference in the slope suggests different carrier densities of the two samples.[18] It appears that after the toluene treatment, the sample becomes more intrinsic (intrinsic semiconductor means a pure

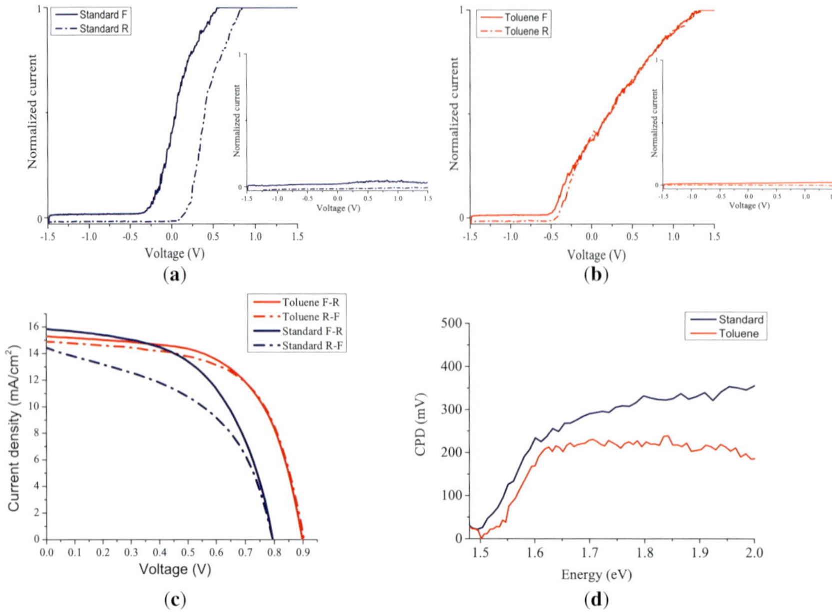

Figure 11. IV plots measured on single perovskite grain using cAFM technique of (**a**) nontreated perovskite film, and of (**b**) toluene-treated perovskite film. (**c**) JV curves measured by solar simulator for HCL-free PSCs. The scan rate was 0.087 V/s. (**d**) SPV measurements of toluene-treated and nontreated perovskite films, the black arrow indicates the difference in the ΔCPD. F, forward; R, reverse. Taken from Ref. 3 with permission of the American Chemical Society.

semiconductor without any significant dopant species present) than without the toluene treatment. (4) In the case of nonconductive grains, the JV plots were almost zero (insets of Figure 11a and b). From the cAFM measurements, it seems that the toluene treatment does not just passivate the perovskite film, but it also changes its electronic properties.

Figure 11c shows JV curves of forward and reverse scans, measured by solar simulator under 1 sun illumination of toluene-treated and nontreated HCL-free cells. The difference in the curve between the toluene-treated cells compared to the nontreated cells is clear; the hysteresis in the nontreated cells is much more pronounced than in the toluene-treated cells where a relatively small change appears between the forward-to-reverse scan and the reverse-to-forward scan. However, in contrast to the JV plot of the toluene-treated film measured by the cAFM, where no hysteresis was observed (Figure 11b), in the current voltage measurements of the complete solar cell (Figure 11c), there is still a small shift between the two scan directions. This important result suggests that the origin of the hysteresis has more than one influence, when clearly one influence on the hysteresis is related to the intrinsic properties of the perovskite, probably the memory effect.[18]

To further elucidate the influence of the toluene treatment on the electronic properties of the perovskite, the SPV technique was applied on the toluene-treated and nontreated perovskite films (Figure 11d). The main observation noted from the SPV spectra is the difference in the contact potential difference (ΔCPD). The ΔCPD is higher by approximately 100 mV (marked with an arrow in Figure 11d) in the case of the nontreated perovskite film compared to the toluene-treated film. The difference in ΔCPD suggests that the quasi-Fermi level of the toluene-treated film is higher (less negative by 0.1 eV) than the quasi-Fermi level of the nontreated toluene film. Therefore, it can be concluded that on subsequent toluene treatment, the perovskite film becomes slightly more intrinsic as also observed by the cAFM measurements. Beyond that, the SPV indicates the approximated band gap energy of the perovskite. The band gap of the toluene-treated film, as extracted from the SPV spectra, is 1.57 eV, while for the nontreated film, the band gap is 1.56 eV, which suggests that there is no observable change in the band gap.

Figure 12 presents a schematic illustration of the effect of suggested toluene treatment on the perovskite surface. During the toluene treatment, excess halides and MA ions are removed from the surface by forming a complex with the toluene, similar to the previous report by Seok and coworkers.[19] This "cleaning" creates a net positive charge on the Pb^{2+} ions. Snaith and coworkers have reported on similar effect before the application of Lewis base passivation.[15]

This interpretation is correlated to the results obtained from the cAFM. The cAFM measurement indicates that the surface after toluene treatment is more conductive than before the toluene treatment, which agrees well with the net positive charge on the Pb^{2+} ions in the case of toluene-treated perovskite surface. In addition, the net positive charge of the perovskite surface after the toluene treatment is beneficial for the PV

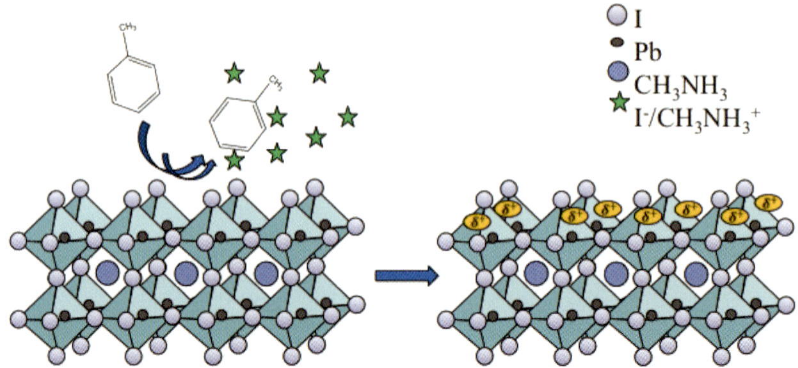

Figure 12.   The effect of the toluene treatment on the perovskite surface. During the toluene treatment, excess of halide and MA ions are removed from the surface which creates a net positive charge on the Pb^{2+} ions. Taken from Ref. 3. with permission of the American Chemical Society.

performance. Net positive charge of the perovskite surface could accept electrons more efficiently, which is useful for the interface of the perovskite with the gold contact.

III. High-Voltage Hole Conducting Layer-Free Solar Cells: The Oxide Role in the PSC Configuration

One of the attractive properties of organometal halide perovskite is its ability to gain high V_{oc} with a high ratio of $V_{oc}/(E_g/q)$. This is the ratio of the maximum voltage developed by the solar cell (V_{oc}) to the voltage related to the band gap of the absorber (E_g/q). Through this section, several combinations of metal oxides/perovskites are studied. As a result of successive combination of perovskite and oxide, high V_{oc} of 1.35 V was achieved in PSCs without HCL. SPV and incident modulated photovoltage spectroscopy (IMPS) are used to elucidate the reason for the high voltage achieved for these HCL-free SCs.[20]

Figure 13a shows a general scheme of the high-voltage HCL-free PSC. The bottom layer is composed of glass substrate coated with TCO and a TiO_2 compact layer. Then a thin film of nanocrystalline metal oxide

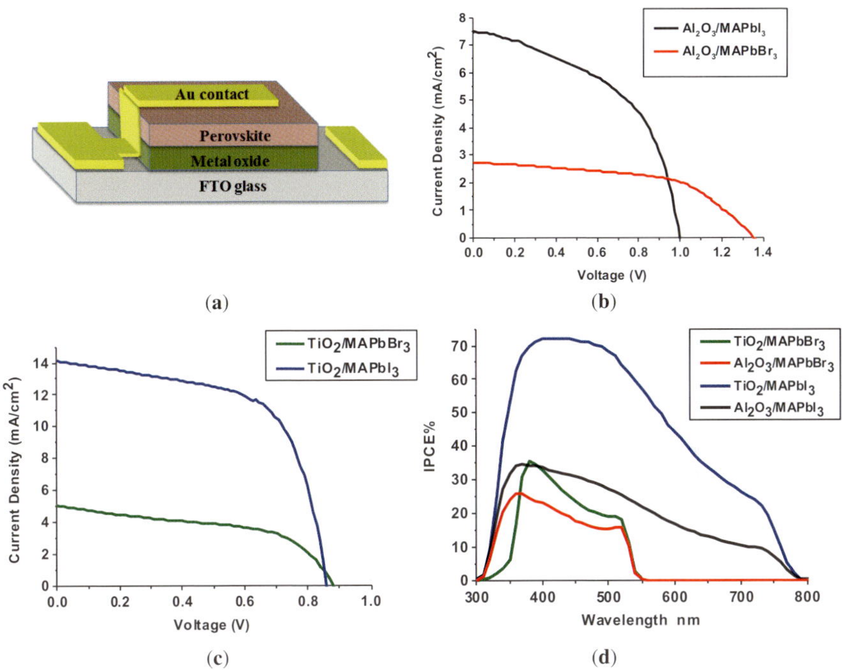

Figure 13. (**a**) The solar cell structure. JV curves of the cells made with (**b**) Al_2O_3 as scaffold and (**c**) TiO_2 as scaffold. (**d**) The corresponding IPCE curves. Taken from Ref. 21 with permission of the Royal Society of Chemistry.

TiO_2 or Al_2O_3 was deposited. The $MAPbI_3$ or $MAPbBr_3$ perovskites were deposited by the two-step deposition as described earlier,[4,14] Followed by a metal contact evaporation on top of the perovskite.

Figure 13b–d and Table 4 present the PV parameters and the IPCE achieved for the high-voltage cells. Four different combinations were studied; nanocrystalline TiO_2 with $MAPbI_3$ and $MAPbBr_3$, and nanocrystalline Al_2O_3 with both perovskites. The V_{oc} for the cells with the $MAPbBr_3$ perovskite deliver higher voltages compared with the cells with the $MAPbI_3$ perovskite related to the same metal oxide. Moreover, the cells with Al_2O_3 achieve higher V_{oc} compared to the TiO_2-based cells. The highest V_{oc} observed for the $Al_2O_3/MAPbBr_3$ configuration achieved 1.35 V without a hole conductor. This is the highest reported open-circuit voltage for perovskite cells without HCL and is comparable to cells using

Table 4. PV parameters of the studied HCL-free SCs. Taken from Ref. 21 with permission of the Royal Society of Chemistry.

	Efficiency (%)	FF	J_{sc} (mA/cm^2)	V_{oc} (V)
TiO$_2$/MAPbI$_3$	7.5	0.61	14.1	0.86
TiO$_2$/MAPbBr$_3$	1.88	0.49	4.37	0.87
Al$_2$O$_3$/MAPbI$_3$	4.13	0.50	7.46	1.00
Al$_2$O$_3$/MAPbBr$_3$	2.02	0.55	2.70	1.35

hole transport material. It is important to note that the average V_{oc} (including more than 10 cells) for the Al$_2$O$_3$/MAPbBr$_3$ configuration is 1.24 V. High PCE with high voltage was observed for the Al$_2$O$_3$/MAPbI$_3$ configuration, which achieved PCE of 4.1% with V_{oc} of 1 V. The IPCE spectra show typical behavior of absorption until 550 nm for the MAPbBr$_3$ and until 780 nm for the MAPbI$_3$.

Figure 14a and b shows the energy level diagram for the four different cases described here. Figure 14a presents the MAPbBr$_3$ and MAPbI$_3$ deposited on Al$_2$O$_3$ which functions as a scaffold only; electron injection from the perovskite to the Al$_2$O$_3$ is not possible in this configuration. Figure 14b presents the MAPbBr$_3$ and MAPbI$_3$ upon TiO$_2$, where electron injection from the perovskite to the TiO$_2$ is favorable. SPV measurements were used to measure the work functions of the Al$_2$O$_3$, TiO$_2$, MAPbI$_3$, and MAPbBr$_3$; the calculated work functions are shown as red lines in Figure 14a and b. The work function positions (which are the Fermi level positions) correspond well to the p-type behavior of the MAPbBr$_3$ and MAPbI$_3$ perovskites, and the n-type behavior of the TiO$_2$, as discussed below. Figure 14c and d shows the corresponding HR-SEM cross-sectional images of the Al$_2$O$_3$/MAPbI$_3$ and Al$_2$O$_3$/MAPbBr$_3$ HCL-free cells and the perovskite over layer can be observed clearly.

SPV spectroscopy and intensity-modulated photovoltage spectroscopy (IMVS) were performed to gain more information about the reason for the high V_{oc} when no HCL is used. Previous studies already demonstrate high V_{oc} in perovskite cells containing HCL. It was suggested that the V_{oc} is not merely the difference between the hole Fermi level of the hole conductor and the electron Fermi level of the nanocrystalline TiO$_2$.[13,22] Moreover, it was reported that charges could be accumulated in the perovskite due to its high capacitance, which allows the control of the quasi Fermi level during illumination.[23]

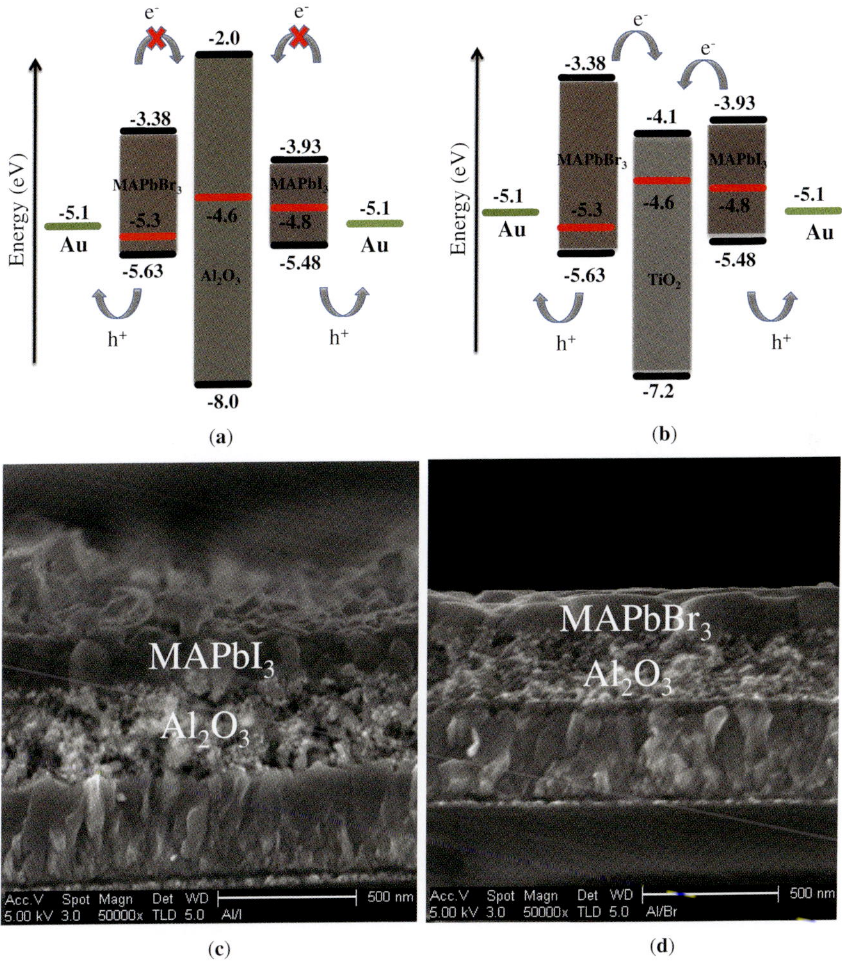

Figure 14. (**a** and **b**) Energy level diagram of the different cells. Fermi levels measured under dark are presented in red in the figure. The position of the conduction and valence bands was taken from literature.[21] HR-SEM cross section of (**c**) Al_2O_3/MAPbI$_3$ HCL-free cell and (**d**) Al_2O_3/MAPbBr$_3$ HCL-free cell. Taken from Ref. 21 with permission of the Royal Society of Chemistry.

The SPV spectra of the MAPbI$_3$ and MAPbBr$_3$ are shown in Figure 15a, with the estimated band gaps shown as vertical lines. Several observations result from the SPV spectra (Figure 15a). First, the spectra provide information regarding the band gaps of the materials, equivalent to the information observed from the absorption. Second, the sign of the

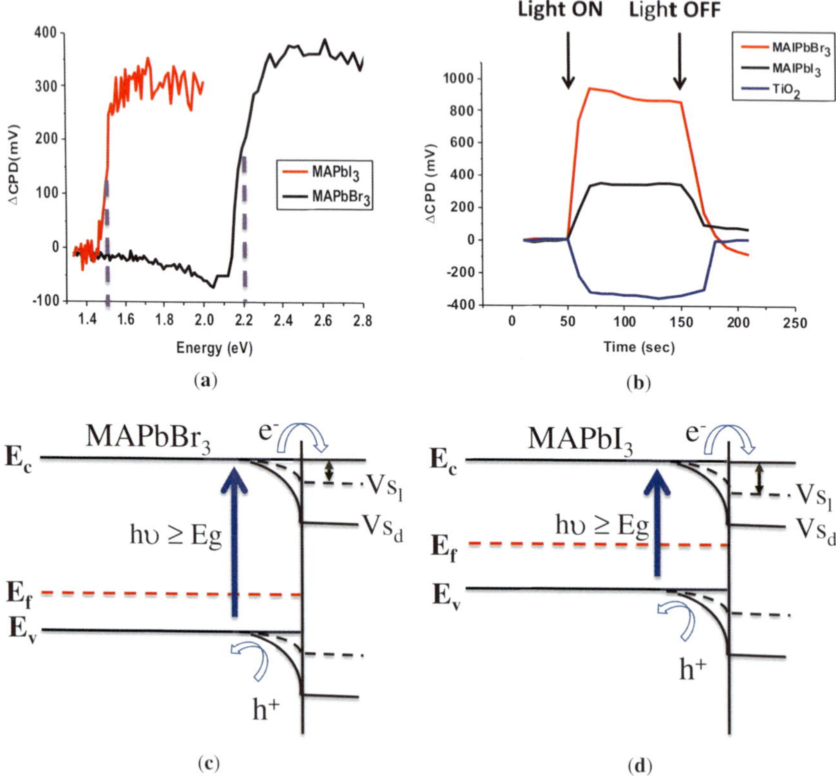

Figure 15. (a) SPV spectra of the MAPbI$_3$ and MAPbBr$_3$ films with estimated band gaps, 1.55 and 2.2 eV, respectively. (**b**) CPD change with white light switched on and off for the samples studied. (**c**) The effect of band-to-band transitions on the SPV responses of MAPbBr$_3$ and (**d**) MAPbI$_3$. V_{sd} is the surface potential in the dark and V_{sl} is the surface potential in light. Taken from Ref. 21 with permission of the Royal Society of Chemistry.

SPV signal indicates the samples type. The surface work function is changed under illumination; it decreases for the n-type semiconductor, TiO$_2$ in this case, and increases for the p-type semiconductor, the MAPbI$_3$ and MAPbBr$_3$ in this case. A third important observation is related to the unique property of the SPV method, which is its immunity to reflection or scattering losses only photons that were absorbed in the sample contribute to the SPV signal. In this case, the signal onset starts at photon energies very close to the band gap of the perovskite samples (both $MAPbI_3$ and $MAPbBr_3$). As a result, it can be concluded that the perovskite samples have only few sub-band gap states. The calculation of the $V_{oc}/(E_g/q)$ fraction for the $Al_2O_3/MAPbBr_3$ cell, gives the result of 0.61, compared to recent reports of high-voltage perovskite cells with HCL which presented $V_{oc}/(E_g/q)$ of 0.55 and 0.73.[24,25] These results demonstrate comparable values with respect to the values obtained with HCL. It should be noted that the relation $V_{oc}/(E_g/q)$ is 0.64 for the $Al_2O_3/MAPbI_3$ cell, which suggests that it has slightly less thermal losses than the $Al_2O_3/MAPbBr_3$ cell.

The $V_{oc}/(E_g/q)$ relation in the case of Al_2O_3-based cells are 0.64 and 0.61 for the $MAPbI_3$ and $MAPbBr_3$, respectively, whereas for the TiO_2-based cells this calculation gives values of 0.55 and 0.37 for $MAPbI_3$ and $MAPbBr_3$, respectively. Based on these values, it can be observed that the $Al_2O_3/MAPbBr_3$-based cells have slightly more thermal losses than the $Al_2O_3/MAPbI_3$ cells. Moreover, it is clear that dominant thermal losses were observed in the TiO_2-based cell compared to the Al_2O_3-based cells.

Figure 15b presents the CPD change when a light is switched on and off. The SPV onsets, t_{on} and t_{off}, are below the resolution limit of the measurements system. Observations from these measurements are related to the change in the CPD; for the TiO_2 sample a negative change in the CPD was observed, while for the perovskites samples a positive change in the CPD was observed, corresponding to their electronic behavior. The ΔCPD for the $MAPbI_3$ is 350 mV and the ΔCPD for the $MAPbBr_3$ is 850 mV.

The V_S, which is presented in Figure 15c and d, is the surface potential barrier, which was measured during the SPV measurements. The difference between the surface potential in the light (V_{sl}) and in the dark (V_{sd}) is defined as the SPV signal. In super band gap illumination, photons with energy equal or larger than the band gap hit the material and generate electron-hole pairs, which are collected by the surface barrier. Consequently, the surface potential is reduced. The trap-to-band transition is neglected in super band gap illumination while the band-to-band absorption is the dominant one.

Figure 15c and d shows the effect of band-to-band transition on the SPV response of $MAPbBr_3$ and $MAPbI_3$ respectively. Under illumination there is redistribution of surface charges, which decrease the band

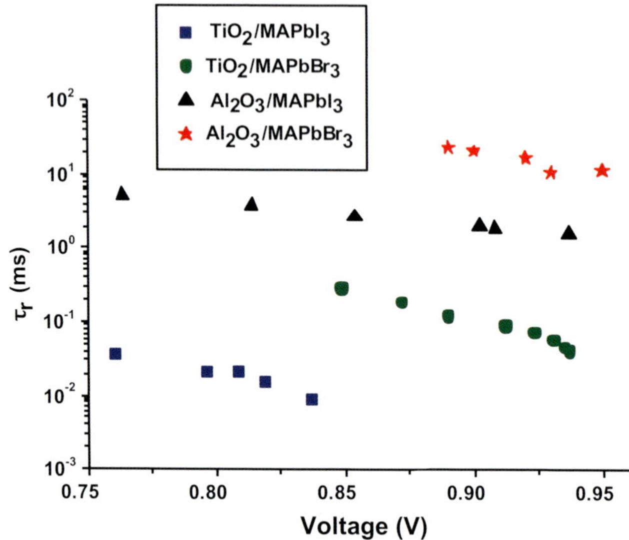

Figure 16.   Recombination lifetime (τ_r) as a function of the open-circuit voltage for the cells studied, measured by IMVS. Taken from Ref. 21 with permission of the Royal Society of Chemistry.

bending, and as a result the SPV response is generated. According to the ΔCPD shown in Figure 1b, the surface potential for the MAPbBr$_3$ is smaller than the surface potential for the MAPbI$_3$ as indicated by the bidirectional arrow in Figure 15c and d respectively (this is also seen by the band bending reduction, since the difference between V_{sl} to V_{sd} is larger in the case of MAPbBr$_3$). The reduction of the surface potential, which is observed from the SPV measurements for the HCL-free MAPbBr$_3$ cells, could possibly contribute to the higher V_{oc} achieved for these cells.

Further contribution for the difference in the V_{oc} is presented in Figure 16. The recombination lifetime (τr) as a function of the voltage were calculated by IMVS.[26–28] All cells showed similar dependence of decrease in τr with increasing the voltage. This behavior can be attributed to the increased recombination with the higher electron density.[26] However, the τr values are different for the various cells; in particular, low τr values were observed for the TiO$_2$/MAPbI$_3$ cell, which also had relatively low V_{oc}. The highest τr values were observed for the Al$_2$O$_3$/MAPbBr$_3$ cell.

This reflects that in this sample the recombination processes were more rare (since the lifetime for recombination is longer), and this corresponds the high V_{oc} which was recorded. Longer recombination lifetime decrease recombination processes, which could result in higher V_{oc} values.[29,30] In addition, longer recombination lifetimes were observed for the cells with the Al_2O_3 scaffold, compared to cells with mesoporous TiO_2. This could also contribute to the higher V_{oc} values which were observed in the case of Al_2O_3-based cells. This is consistent with the qV_{oc}/Eg relation as discussed earlier; more thermal losses were observed for TiO_2-based cells as the metal oxide compare to Al_2O_3-based cells.

IV. Planar Hole Conducting Layer-Free Perovskite Solar Cell

In this section, a unique planar HCL-free PSC is discussed. In contrary to the previous sections, in this part, the perovskite deposition was done by a facile spray technique. This approach led to the formation of perovskite film that consists of micron-sized grains. This deposition technique was used for the fabrication of simpler solar cell configuration, consisting of glass/TCO/compact TiO_2/MAPbI$_3$/Au. This means that the perovskite functioned both as a light harvester and a hole conductor in this SC structure. The perovskite film thickness and the blocking layer thickness were varied. Interestingly, PCE of 6.9% was achieved for the planar HTM-free cell with 3.4 nm perovskite film thickness. Capacitance voltage measurements shed more light on the operation mechanism of this unique configuration.[31]

Figure 17a shows a schematic illustration of the spray deposition technique. The perovskite precursors are sprayed from a one-step solution onto a hot substrate. Subsequently, the solvent [N, N-dimethylformamide, (DMF)] rapidly evaporates, and the perovskite crystals are immediately created. Rapid evaporation of the solvent, which prevents percolation of the perovskite crystals into a mesoporous film, is an important stage in the deposition technique. Thus, when using this deposition technique, low PV performance is expected to be obtained in a mesoporous configuration. Consequently, the facile deposition technique was used in a planar architecture as shown schematically in Figure 17b. As confirmed by the XRD,

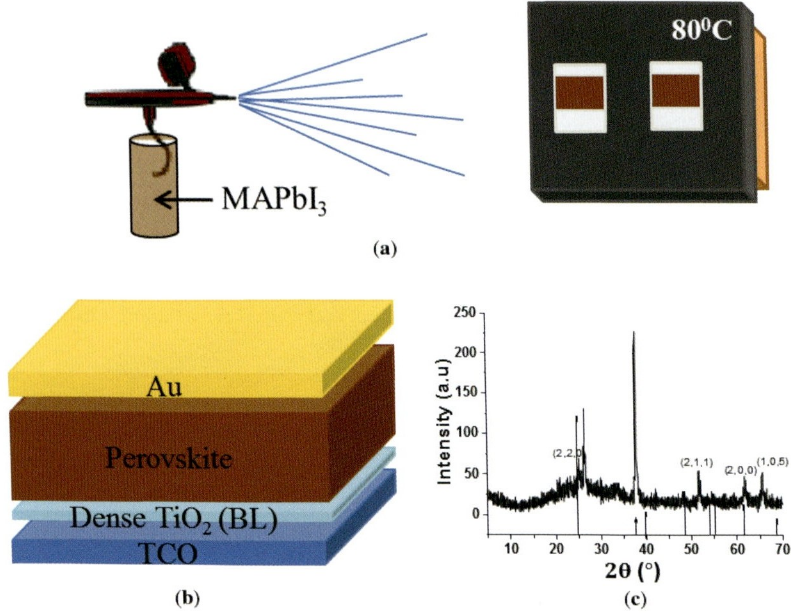

Figure 17. (**a**) Schematic illustration of the spray deposition technique. (**b**) Planar HCL-free PSC structure. (**c**) XRD of the planar TiO_2 thin film. The anatase TiO_2 crystallographic planes are indexed in the figure. Taken from Ref. 32 with permission of the American Chemical Society.

the dense TiO_2 film reflects anatase peaks (Figure 17c). A thick perovskite film was deposited on the top of the planar TiO_2 layer by spray deposition, which was followed by evaporation of gold back contact. The number of spray passes made over the electrode controlled the thickness of the perovskite film. In this SC structure, the perovskite functions as both a light harvester and a hole conductor which makes this solar cell configuration one of the simplest PV cell structures (no electron conducting layer (ECL) and HCL are used).

One of the exceptional properties of this technique is the ability to create thick perovskite films depending on the number of spray passes. The representative absorption spectrum which is presented in Figure 18a shows the absorption of perovskite layer that was acquired after 10 passes of the spray deposition. The onset of the absorption spectra shows a red shift relatively to the usual absorption spectra of $MAPbI_3$; this

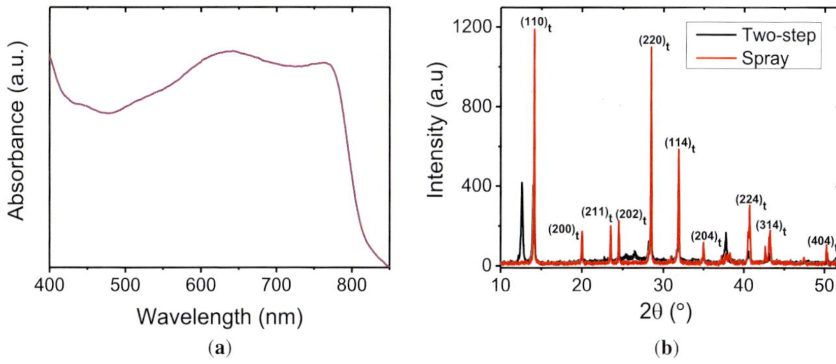

Figure 18. (a) Absorption spectra of perovskite layer that was acquired after 10 passes of the spray deposition. (b) XRD of MAPbI$_3$ made by two-step technique and by spray technique (t, tetragonal). Taken from Ref. 32 with permission of the American Chemical Society.

phenomenon was reported earlier by Grancini *et al.*[32] It was observed that the deposition of MAPbI$_3$ on a flat surface could cause a red shift of the absorption spectra compared to mesoporous film. This red shift is probably due to the slow rotation of the methylammonium cation in the planar structure, which screens the excitonic transition at the onset of the absorbance spectra.[32] In addition, an excitonic feature was also observed in the absorbance spectra for the perovskite film deposited on a flat substrate.[33]

Figure 18b shows the XRD of the MAPbI$_3$ perovskite film that was deposited by two-step deposition technique and by spray deposition technique.[14] As indicated in the figure, the MAPbI$_3$ peaks are clearly seen. However, two main differences are observed between these deposition techniques: (1) The XRD peaks of the spray deposition technique are much stronger, suggesting better orientation of the MAPbI$_3$ grains with the spray technique. (2) The PbI$_2$ peak doesn't appear in the spray deposition technique, in contrast to the two-step deposition, suggesting a complete conversion of the precursors to MAPbI$_3$ perovskite.

Cross-sectional HR-SEM images are shown in Figure 19 for the four different perovskite thicknesses made by 4, 6, 8, and 10 spray passes. Figure 19 inset shows that the spray deposition technique creates micron-sized grains of perovskite depending on the number of spray passes

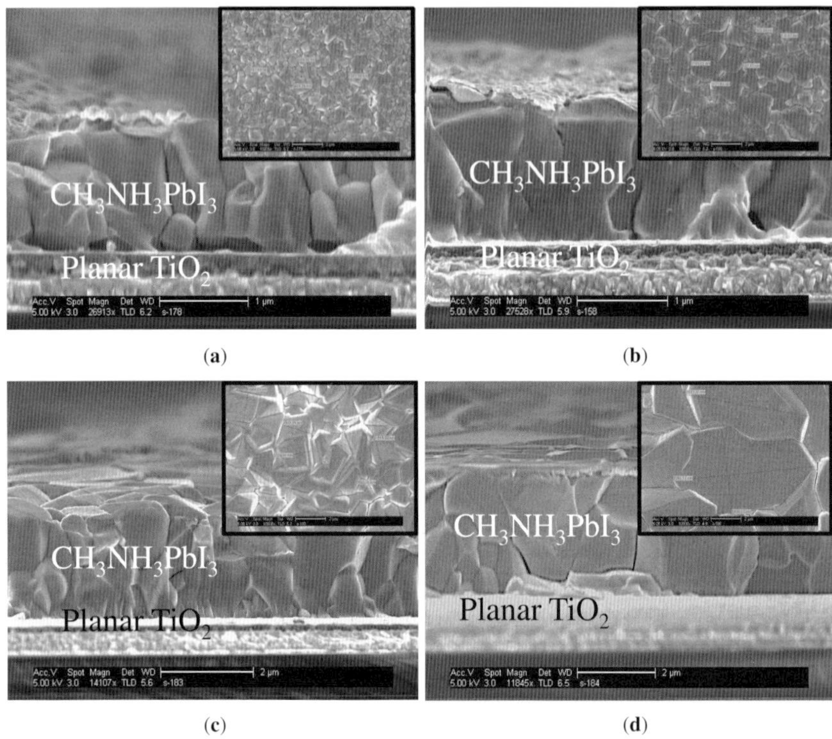

(a)

(b)

(c)

(d)

Figure 19. (a) Four passes of perovskite spray; (b) six passes of perovskite spray; (c) eight passes of perovskite spray; and (d) 10 passes of perovskite spray. Inset: Top view of different spray passes 4, 6, 8, and 10 respectively. Scale bar at the insets is 2 nm. Taken from Ref. 32 with permission of the American Chemical Society.

(discussed in further detail), thus minimizing the grain boundaries, which is beneficial for the PV performance.

To investigate the PV performance of these cells, the thickness of the dense TiO_2 films, was changed and the number of deposition repetitions was varied. Table 5 shows the thicknesses of the dense TiO_2 BL films as measured by profilometer. Clearly, as the number of BL deposition repetitions increases, the thickness of the BL increases. The PV results of cells made using different BL thicknesses (with a constant number of 10 perovskite spray passes) is presented in Table 5 and Figure 20a. The best PV performance was achieved for three BL deposition repetitions, with V_{oc} of

Table 5. PV parameters of cells according to various thicknesses of dense TiO_2 layers (blocking layers); thicknesses as listed in the table. The number of spray passes is constant, equal to 10. Taken from Ref. 32 with permission of the American Chemical Society.

Number of deposition repetitions	V_{oc} (V)	J_{sc} (mA/cm^2)	FF	Efficiency (%)	Dense TiO_2 thickness (nm)
1	0.55	9.35	0.39	1.99	74.4 ± 11
2	0.66	21.54	0.32	4.63	76.2 ± 15
3	0.69	23.01	0.43	6.93	166.9 ± 19
4	0.65	21.64	0.40	5.58	184.2 ± 13
5	0.68	17.82	0.42	5.11	227.5 ± 42

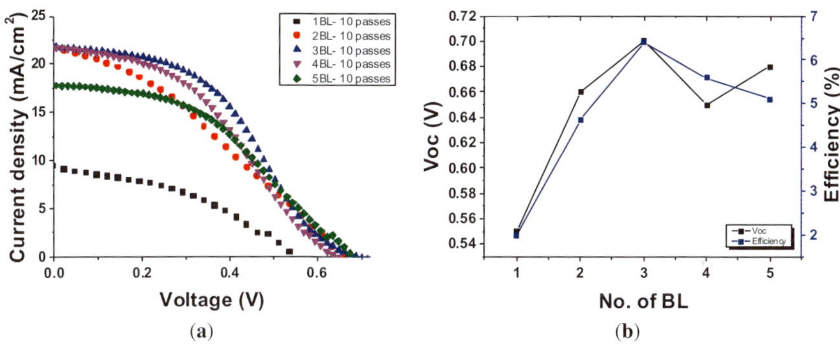

Figure 20.   (**a**) JV curves of cells made by 10 passes deposited on various BL; (**b**) V_{oc} and efficiency as a function of the number of BL. Taken from Ref. 32 with permission of the American Chemical Society.

0.69 V, J_{sc} of 23.01 mA/cm^2, and PCE of 6.9% (Figure 20b). It can be observed that the V_{oc} is barely changed when alternating the BL thickness, (excluding one blocking layer, discussed below). This suggests that from the V_{oc} point of view, the most influencing factor is the perovskite film thickness (the number of spray passes). The s-shape observed in the JV curves of all cells might be related to charge accumulation, discussed below.

The planar HCL-free SCs including 3 BLs showed the best PV performance. In the case of 1 and 2 BLs, the dense TiO_2 was too thin and

Table 6. PV parameters of the cells made upon 3 BLs according to the number of spray passes. The MAPbI$_3$ thickness was extracted from the HR-SEM images shown in Figure 19. Taken from Ref. 32 with permission of the American Chemical Society.

Number of passes	V_{oc} (V)	J_{sc} (mA/cm^2)	FF (%)	Efficiency (%)	Thickness of CH$_3$NH$_3$PbI$_3$ (μm)
4	0.49	13.18	39.6	2.54	1.40
6	0.50	19.49	28	2.74	1.51
8	0.61	17.01	31.6	3.28	2.13
10	0.69	23.01	43.4	6.93	3.37

some pinholes were observed in the blocking layer. On the other hand, in the case of 4 and 5 BLs, it seems that the dense TiO$_2$ is too thick for electrons to transfer efficiently (also calculated below by the depletion region); therefore, the efficiency is decreased.

To optimize the PV performance, the number of spray passes was changed (while using three layers of dense TiO$_2$) as shown in Table 6 and Figure 21a. It can be seen that the efficiency is highest in the case of 10 spray passes as shown in Figure 21b. The reason for the best PV performance in the case of 10 spray passes is mainly due to the difference in the perovskite crystal size. As indicated by the HR-SEM images in Figure 19, the perovskite crystal size increases with the number of spray passes. In the rest of the discussion, 12 and 14 spray passes were excluded since the PV performance decreased significantly in these cases (PCE of 3.1 and 2.9% of 12 and 14 spray passes, respectively). The reason for the decrease in performance for the 12 and 14 passes is the fact that the perovskite layer becomes very thick and as a result it is peeled from the BL, thus harming the PV performance. Figure 21c shows the change in the perovskite crystal size as a function of number of spray passes. In the case of 10 spray passes, the crystal size of the perovskite is 7.5 ± 1 nm (though, a few crystals were smaller than 6.5 nm and some were larger than 8.5 nm). Nie et $al.$[34] have reported that large perovskite grains have less defects and higher mobility enabling better charge carrier transport through the perovskite film. Figure 21a–c supports this argument; the V_{oc} and the efficiency increase with the number of spray passes (bigger perovskite crystals). The increased number of spray passes points out the

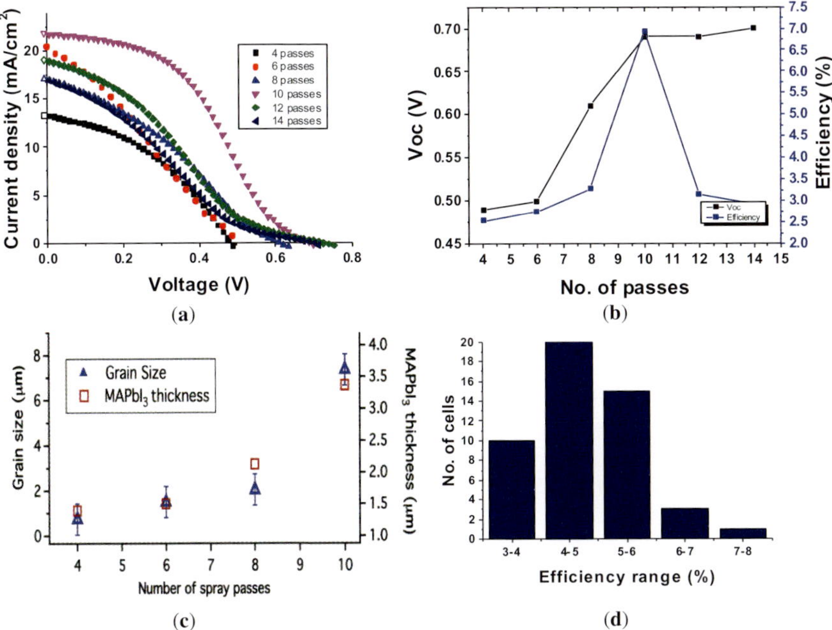

Figure 21. (**a**) JV curves of the cells made on 3 BLs according to the number of spray passes. (**b**) V_{oc} and efficiency as a function of the number of passes. (**c**) Average grain size and the MAPbI$_3$ film thickness as a function of the number of spray passes. (**d**) Statistics of the HCL-free planar solar cells. Taken from Ref. 32 with permission of the American Chemical Society.

decrease in recombination and defects in the perovskite crystals. Moreover, as indicated by the top-view SEM images (not shown), the coverage is improved when increasing the number of spray passes, which further supports the increase in the efficiency.

Statistics for the planar HCL-free SCs are presented in Figure 21c. An average efficiency of 4.7 ± 1% was observed for over 50 cells.

To elucidate the mechanism of these planar HCL-free PSCs, CV measurements were performed under both dark and illumination. Figure 22b shows the results of CV measurements for the cell includes 3 BLs and 10 spray passes of perovskite (this cell demonstrated the best PV performance). All other cells discussed in this manuscript demonstrated the same CV behavior as the cell shown in Figure 22b. In the dark,

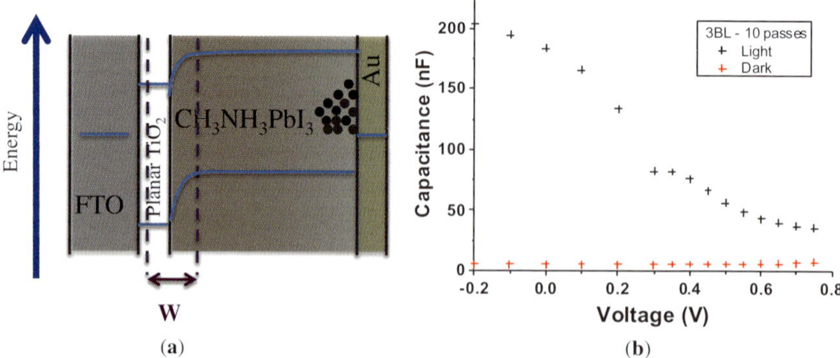

Figure 22. (a) Schematic illustration of the energy level diagram of the planar HCL-free PSC. (b) Capacitance voltage measurements in the dark and under illumination of the cell with 3 BLs and 10 perovskite spray passes. Taken from Ref. 32 with permission of the American Chemical Society.

the capacitance remained constant at approximately 5×10^{-9} F. The capacitance under illumination is two orders of magnitude larger than the capacitance in the dark. Under illumination, the capacitance started to increase, approaching the V_{oc} of the cell. This implies that photogenerated carriers are accumulated within the solar cell and can't be effectively collected by the electrode as shown schematically in Figure 22a. Since the perovskite thickness in this planar structure is in the range of 1.4–3.4 μm, which is much larger than the perovskite optimum thickness according to its absorption coefficient,[35] charges are accumulated at the interface with the metal contact (supported by the CV measurements). The accumulation of charges might be the explanation to the S-shape observed in the JV curves of these cells. The S-shaped curves contribute to the low FF of the planar HCL-free solar cells. It is important to indicate that the observed S-shape was independent on the JV scan rate.

From the discussion presented, it could be construed that thinner perovskite film would be beneficial for the planar HTM-free cell. According to Table 6, the PCE at four passes is much lower than the PCE at 10 passes (which produced thicker and larger perovskite crystals as mentioned). Therefore, it could be argued that besides the charge accumulation at the perovskite/metal interface, there is an additional contribution to the operation mechanism of the planar HCL-free cell.

Mott Schottky analysis was performed on these solar cells to investigate the blocking layer (dense TiO_2)\perovskite interface and to extract the depletion region width.[6]

Using eq 2, the widths of depletion regions at the planar TiO_2/perovskite junctions for 3, 4, and 5 BLs were calculated; where W_n is the depletion through the anatase TiO_2 side, W_p is the depletion through the perovskite side, and W_t is the total depletion region. In the case of 1 and 2 BLs, the PV performance was low; therefore, the Mott Schottky analysis was performed for the other BL thicknesses only, while maintaining 10 passes of the perovskite spray deposition. The results show similar W_t for the three cases: $W_t = 267$ nm for 3 BL with W_n, W_p of 150 nm, 117 nm; $W_t = 239$ nm for 4 BL with W_n, W_p of 112 nm, 127 nm; $W_t = 238$ nm for 5 BL with W_n, W_p of 121 nm, 127 nm, respectively.

The depletion region width for the 3 BL is slightly wider than for the other samples. However, the results for the perovskite films ($Wp = 117–127$ nm) indicate that most of the perovskite film is not depleted. Under illumination close to the junction, the carriers — the holes in this case — are transported to the back contact and cross the whole perovskite film. It is possible that the long diffusion lengths of electrons and holes are not solely responsible for the operation of this planar HCL-free cell. It is suggested that once a charge separation occurs close to the depletion region, electrons drift to the edge of the depletion region (close to the TCO) while holes are transported to the back contact. In the case of thick perovskite film, the holes are transported without any interference through the perovskite film, because far away from the depletion region there are no free electrons and holes that can recombine with the transported holes. On the other hand, if the charge separation occurs far from the dense TiO2/ perovskite junction, a recombination will probably occur. This is probably one of the reasons for the low V_{oc} observed in these cells.

V. Conclusions

In this chapter, the parameters which influence the PV performance of HCL-free PSCs were discussed. Optimization of the two-step deposition technique and the addition of antisolvent treatment to the HCL-free PSCs were presented. It was observed that the antisolvent treatment improves not just the surface morphology and coverage but also influences the electronic properties of the perovskite film and reduce the hysteresis effect. A good correlation was found between the depletion region width at the TiO_2/$MAPbI_3$ junction and the measured PCE which shed light on the mechanism of these solar cells.

High open-circuit voltage of 1.35 V was observed for HCL-free PSCs. The position of the Fermi level and the SPV spectra of $MAPbI_3$ and $MAPbBr_3$ revealed the p-type nature of these perovskites. The high open-circuit voltage observed in cells without HCL indicates that the V_{oc} is strongly related to the perovskite/metal oxide interfaces. In the last section, planar HCL-free PSCs were discussed. Micron-sized perovskite films were deposited by spray technique which enable facile control over the perovskite thickness which deposited onto compact TiO_2 layer. Two main contributions were recognized for the PV mechanism in this planar HCL-free PSC structure. A depletion region at the planar TiO_2/perovskite junction and charge accumulation at the perovskite/metal oxide interface were recognized. This chapter summarizes some of the recent activity on hole-conductor-free perovskite-based solar cells presenting the potential of using the perovskite in this special configuration.

VI. References

1. Etgar, L.; Gao, P.; Xue, Z.; Peng, Q.; Chandiran, A. K.; Liu, B.; Nazeeruddin, M. K.; Grätzel, M. *J. Am. Chem. Soc.* **2012**, *134*, 17396–17399.
2. Aharon, S.; Gamliel, S.; Cohen, B. E.; Etgar, L. *Phys. Chem. Chem. Phys.* **2014**, *16*, 10512–10518.
3. Cohen, B.-E.; Aharon, S.; Dymshits, A.; Etgar, L. *J. Phys. Chem. C* **2016**, *120*, 142–147.
4. Burschka, J.; Pellet, N.; Moon, S.-J.; Humphry-Baker, R.; Gao, P.; Nazeeruddin, M. K.; Gratzel, M. *Nature* **2013**, *499*, 316–319.
5. Laban, W. A.; Etgar, L. *Energy Environ. Sci.* **2013**, *6*, 3249–3253.
6. Schottky, W. *Z. Phys.* **1942**, *118*, 539–592.
7. Clifford, J. P.; Johnston, K. W.; Levina, L.; Sargent, E. H. *Appl. Phys. Lett.* **2007**, *91*, 253117.
8. Poglitsch, A.; Weber, D. *J. Chem. Phys.* **1987**, *87*, 6373–6378.
9. Baikie, T.; Fang, Y.; Kadro, J. M.; Schreyer, M.; Wei, F.; Mhaisalkar, S. G.; Graetzel, M.; White, T. J. *J. Mater. Chem. A* **2013**, *1*, 5628–5641.
10. Luther, J. M.; Law, M.; Beard, M. C.; Song, Q.; Reese, M. O.; Ellingson, R. J.; Nozik, A. J. *Nano Lett.* **2008**, *8*, 3488–3492.

11. O'Regan, B.; Moser, J.; Anderson, M.; Graetzel, M. *J. Phys. Chem.* **1990**, *94*, 8720–8726.

12. Nakade, S.; Saito, Y.; Kubo, W.; Kanzaki, T.; Kitamura, T.; Wada, Y.; Yanagida, S. *Electrochem. Commun.* **2003**, *5*, 804–808.

13. Zaban, A.; Meier, A.; Gregg, B. A. *J. Phys. Chem. B* **1997**, *101*, 7985–7990.

14. Cohen, B.-E.; Gamliel, S.; Etgar, L. *APL Mater.* **2014**, *2*, 081502.

15. Noel, N. K.; Abate, A.; Stranks, S. D.; Parrott, E. S.; Burlakov, V. M.; Goriely, A.; Snaith, H. J. *ACS Nano* **2014**, *8*, 9815–9821.

16. Luber, E. J.; Buriak, J. M. *ACS Nano* **2013**, *7*, 4708–4714.

17. Mashkoor, A.; Jiong, Z.; Javed, I.; Wei, M.; Lin, X.; Rigen, M.; Jing, Z. *J. Phys. D: Appl. Phys.* **2009**, *42*, 165406.

18. Strukov, D. B.; Snider, G. S.; Stewart, D. R.; Williams, R. S. *Nature* **2008**, *453*, 80–83.

19. Jeon, N. J.; Noh, J. H.; Kim, Y. C.; Yang, W. S.; Ryu, S.; Seok, S. I. *Nat. Mater.* **2014**, *13*, 897–903.

20. Dymshits, A.; Rotem, A.; Etgar, L. *J. Mater. Chem. A* **2014**, *2*, 20776–20781.

21. Ryu, S.; Noh, J. H.; Jeon, N. J.; Chan Kim, Y.; Yang, W. S.; Seo, J.; Seok, S. I. *Energy Environ. Sci.* **2014**, *7*, 2614–2618.

22. Chiang, Y.-F.; Jeng, J.-Y.; Lee, M.-H.; Peng, S.-R.; Chen, P.; Guo, T.-F.; Wen, T.-C.; Hsu, Y.-J.; Hsu, C.-M. *Phys. Chem. Chem. Phys.* **2014**, *16*, 6033–6040.

23. Kim, H.-S.; Mora-Sero, I.; Gonzalez-Pedro, V.; Fabregat-Santiago, F.; Juarez-Perez, E. J.; Park, N.-G.; Bisquert, J. *Nat. Commun.* **2013**, *4*, 2242.

24. Edri, E.; Kirmayer, S.; Kulbak, M.; Hodes, G.; Cahen, D. *J. Phys. Chem. Lett.* **2014**, *5*, 429–433.

25. Edri, E.; Kirmayer, S.; Cahen, D.; Hodes, G. *J. Phys. Chem. Lett.* **2013**, *4*, 897–902.

26. Zhu, K.; Jang, S.-R.; Frank, A. J. *J. Phys. Chem. Lett.* **2011**, *2*, 1070–1076.

27. Zhao, Y.; Zhu, K. *J. Phys. Chem. Lett.* **2013**, *4*, 2880–2884.

28. Zhao, Y.; Nardes, A. M.; Zhu, K. *J. Phys. Chem. Lett.* **2014**, *5*, 490–494.

29. Mozer, A. J.; Wagner, P.; Officer, D. L.; Wallace, G. G.; Campbell, W. M.; Miyashita, M.; Sunahara, K.; Mori, S. *Chem. Commun.* **2008**, 4741–4743.

30. (a) Docampo, P.; Tiwana, P.; Sakai,N.; Miura, H.; Herz, L.; Murakami, T.; Snaith, H. J. *J. Phys. Chem. Lett.* **2014**, *5*, 490; (b) Docampo, P.; Tiwana, P.; Sakai, N.; Miura, H.; Herz, L.; Murakami, T.; Snaith, H. J. *J. Phys. Chem. C* **2012**, *116*, 22840–22846.

31. Gamliel, S.; Dymshits, A.; Aharon, S.; Terkieltaub, E.; Etgar, L. *J. Phys. Chem. C* **2015**, *119*, 19722–19728.

32. Grancini, G.; Marras, S.; Prato, M.; Giannini, C.; Quarti, C.; De Angelis, F.; De Bastiani, M.; Eperon, G. E.; Snaith, H. J.; Manna, L.; Petrozza, A. *J. Phys. Chem. Lett.* **2014**, *5*, 3836–3842.

33. D'Innocenzo, V.; Grancini, G.; Alcocer, M. J. P.; Kandada, A. R. S.; Stranks, S. D.; Lee, M. M.; Lanzani, G.; Snaith, H. J.; Petrozza, A. *Nat. Commun.* **2014**, *5*, 3586.

34. Nie, W.; Tsai, H.; Asadpour, R.; Blancon, J.-C.; Neukirch, A. J.; Gupta, G.; Crochet, J. J.; Chhowalla, M.; Tretiak, S.; Alam, M. A.; Wang, H.-L.; Mohite, A. D. *Science* **2015**, *347*, 522.

35. De Wolf, S.; Holovsky, J.; Moon, S.-J.; Löper, P.; Niesen, B.; Ledinsky, M.; Haug, F.-J.; Yum, J.-H.; Ballif, C. *J. Phys. Chem. Lett.* **2014**, *5*, 1035–1039.

6 Stability Issues of Inorganic/Organic Hybrid Lead Perovskite Solar Cells

Dan Li and Mingkui Wang*

Wuhan National Laboratory for Optoelectronics,
Huazhong University of Science and Technology,
Wuhan, Hubei 430074, China
*mingkui.wang@mail.hust.edu.cn

Table of Contents

List of Abbreviations

Li-TFSI	bis(trifluoromethane)sulfonimide lithium salt
BCP	bathocuproine
DSSC	dye-sensitized solar cells
EA	$C_2H_5NH_3^+$
FA	$HC(NH_2)_2^+$
FAPbI$_3$	formamidinium lead iodide
FTIR	fourier transform infrared spectroscopy
HOMO	highest occupied molecular orbital
HTM	hole transport material
ICBA	indene-C_{60} bisadduct
IPCE	incident photon-to-current conversion efficiency
ITO	indium-tin oxide
LUMO	lowest occupied molecular orbital
MA	$CH_3NH_3^+$
MAPbI$_3$	methylammonium lead iodide
OPV	organic photovoltaic
PCBM	(6,6)-phenyl C_{61}-butyric acid methyl ester
PCE	power conversion efficiency
PEDOT:PSS	poly(3,4-ethylenedioxythiophene) polystyrene sulphonate
PEIE	polyethyleneimine ethoxylated
PHJ	planar heterojunction
PMMA	poly(methylmethacrylate)
PSC	perovskite solar cells
PTAA	poly-(triarylamine)
PV	photovoltaic
spiro-OMeTAD	2,2´,7,7´-tetrakis-(N,N-di-p-methoxyphenylamine) 9,9´-spirobifluorene
SWNTs	single-walled carbon nanotubes
TBP	4-tert-butylpyridine
XPS	X-ray photoelectron spectroscopy
XRD	X-ray diffraction

I. Introduction

In the 21st century, the energy sector will face an increasingly complex array of interlocking challenges—economic, geopolitical, technological, and environmental. As the population continues to expand, the energy needs of billions of additional people in rural and especially in urban areas will have to be met. Meanwhile, supplies of conventional oil and conventional natural gas are expected to decline in the not-too-distant future due to continuous utilization. Meeting these challenges will require very long lead times.

Solar energy in all its facets raises the greatest public expectations. Currently, the conversion of sunlight into energy can be achieved through solar photovoltaic (PV) cells, concentrated PV and solar thermal technologies. On a local basis and for individual countries, solar energy no doubt represents an extremely promising and sustainable method of energy production. PV production of electricity yields less energy per unit area than solar heat collectors, but it is more versatile.

PV is are a promising technology that directly takes advantage of Earth's ultimate source of power, that is, the sun. When exposed to light, solar cells are capable of producing electricity or chemical fuels without any harmful effect to the environment, which means they can generate power for many years while requiring only minimal maintenance and operational costs. Presently, PV provides a relatively small scale of the total global electricity generation (~1%) but the use of solar PV is expanding rapidly due to annual cost decrease of this technology. Currently, the widespread use of PVs over other energy sources is limited by the relatively high cost and low efficiency of solar cells. Solar cells made of crystalline silicon are often referred as first-generation solar cells, which have dominated the PV market over the past half-century. The "third-generation" solar cell technologies have been developed to pursue high power conversion efficiency (PCE) and low cost, including light-concentrator cells, organic photovoltaic (OPV), dye-sensitized solar cells (DSSC), organic–inorganic hybrid perovskite solar cells (PSC), and so on.[1]

Since their first application as visible light absorbers in DSSCs in 2009,[2] the inorganic–organic hybrid lead perovskite materials have evoked a tide of research in this field.[3–8] Only in six-years, the PCE of PSC has explosively increased to 22.1% from its initial value of 3.8%.[2] The achievement obtained in PSCs should be attributed to the unique and excellent properties for the inorganic–organic hybrid lead perovskite materials, including suitable band gap, high absorption coefficient, long charge carrier lifetime, and diffusion length.[4,5,9,10] Inorganic–organic hybrid lead perovskite materials have the generic chemical formula of AMX_3 that has a cubic unit cell.[11] A represents an organic cation, which typically is $CH_3NH_3^+$ (MA), $C_2H_5NH_3^+$ (EA), or $HC(NH_2)_2^+$ (FA), located at the corners of the cubic unit cell.[11–14] M is a divalent metal (Pb^{2+}, Sn^{2+}) residing at the body center.[15–18] X is an halogen anion (Cl^-, Br^-, I^-) locating at the face centers.[19,20] Among various perovskite compositions, methylammonium lead iodide ($MAPbI_3$) has been the prototypical and most researched compound.

Up to now, various device architectures have been developed for PSCs with which excellent PV performance can be achieved according to literature reports. In 2009, Miyasaka and coworkers provided strong evidence that the inorganic–organic lead perovskite can be used as light absorber for DSSCs, resulting in a 3.8% efficient $MAPbI_3$-based device. They also reported a high photovoltage of 0.96 V when $MAPbBr_3$ was used.[2] The devices used the conventional dye-sensitized mesoporous n-type titanium oxide TiO_2 electrode and liquid electrolytes (typically the I^-/I_3^- redox couple). By optimizing the mesoporous TiO_2 film thickness and the loading of perovskite materials, Park and coworkers increased perovskite-sensitized solar cells' PCE to 6.54% in 2011.[21] A 3.6-μm-thick TiO_2 film was firstly modified by $Pb(NO_3)_2$, followed by deposition of ca. 2- to 3-nm-sized $MAPbI_3$ quantum dots. The stability of these devices was poor in a liquid electrolyte cell configuration. However, the high absorption coefficient renders them as one of the best absorbers in solid-state solar cells. In 2012, Park and Grätzel *et al.* demonstrated an excellent work by adopting small molecular 2,2′,7,7′-tetrakis-(N,N-di-p-methoxyphenylamine)9,9′-spirobifluorene (spiro-OMeTAD) as the solid-state hole transport material (HTM) to replace the liquid electrolyte, which dissolves the perovskite and causes the device instability.[22] At the same time, Snaith and coworkers reported similar work in which the insulting Al_2O_3 was used as mesoporous scaffold and spiro-OMeTAD as the HTM, exhibiting a PCE of 10.9%.[23] Since then, this kind of device was called as solid-state mesoporous architecture, in which the perovskite absorber was sandwiched between the anode TiO_2 compact layer/ mesoporous layer and spiro-OMeTAD layer. The application of spiro-OMeTAD dramatically increases both the device efficiency and device stability. The device performance has been improved by controlling of perovskite morphology. For example, the simple and popular one-step spin-coating method has been widely utilized to deposit perovskite from a single precursor solution. $MAPbI_3$ precursor solution can be obtained by dissolving stoichiometric quantities of MAI and PbI_2 in polar solvents such as GBL or DMF.[22,24] Through optimizing the precursor concentration and spin-coating conditions, $MAPbI_3$ films can be formed within the pores of the TiO_2 mesoporous layers. However, the infiltration of mesoporous layers would depend critically on the solution concentration, spin-coating speed and the solvent utilized, which was similar to the deposition of HTMs within the mesoporous TiO_2 layers of a solid-state DSSC.[25] As a result, the crystallization tendency of the perovskite could lead to rough surfaces that could introduce shunt pathways into the solar cells.[24] Later on, Grätzel and coworkers introduced a two-step sequential deposition method to efficiently form the perovskite pigment, which led to a PCE of 15% for the solid-state mesoscopic solar cells and increased batch-to-batch reproducibility.[26] This method involves spin-coating step of PbI_2 solution and dipping step in MAI solution of the as-prepared PbI_2 film. The solar cell efficiency was further pushed to 17% by optimizing the perovskite deposition method, that is, two-step spin-coating procedure, which involves two spin-coating steps of PbI_2 solution and MAI solution.[27] The cuboid size of $MAPbI_3$ was easily controlled by varying the concentration of MAI solution, which was believed to be virtual to device performance. In 2015, Seok and coworkers reported on PSCs with efficiency over 20% by introducing narrow band gap formamidinium lead iodide ($FAPbI_3$) perovskite as light absorber, which has a broader absorption spectrum compared with conventional studied $MAPbI_3$.[28] Grätzel and Hagfeldt have claimed a new world record which was as high as 22.1%, updating on the NREL efficiency chart.[1]

During this period, another device configuration, named planar heterojunction (PHJ) PSC, has been also achieved significant progress.

Snaith and coworkers firstly adapted insulating mesoporous Al_2O_3 as scaffold to replace the mesoporous n-type TiO_2 layer. This led a PCE of 10.9% with increased open-circuit voltage.[23] The discovery of bipolar conductivity and long electron-hole diffusion length has further revealed that mesoporous TiO_2 layer (i.e., electron transport layer ETL) is unnecessary, which has evidenced the deployment of PHJ PSCs.[3,24,29] This kind of architecture excludes the use of mesoporous scaffold and simplifies the device processing route. Therefore, the perovskite morphology is more critical to the device performance, which can be influenced by underlying substrate, annealing temperature, and time.[3] A device with PCE of 15.4% was further achieved by a much simpler approach[30] by vapor deposition of perovskite layer, yielding a short-circuit photocurrent of 21.5 mA·cm^{-2}, an open-circuit voltage of 1.07 V and a fill factor of 0.68. Meanwhile, interface engineering has been provided as another efficient way to enhance the device performance.[31-34] By dopping the TiO_2 with yttrium and modifying indium-tin oxide (ITO) with polyethyleneimine ethoxylated (PEIE), Yang and coworkers reported PHJ pervoskite device with a PCE of 19.3%.[34] The modification of TiO_2 and ITO electrodes achieves favorable energy alignment and facilitates efficient electron transport between ITO and perovskite layers, which suppresses excessive interface recombination and thus enhances device performance.

Meanwhile, the inverted PHJ PSC with similar device structure to organic polymer solar cells has attracted growing attention. In 2013, Chen and coworkers firstly employed perovskite in a device architecture of poly(3,4-ethylenedioxythiophene) polystyrene sulphonate (PEDOT:PSS)/ $MAPbI_3$/(6,6)-phenyl C_{61}-butyric acid methyl ester (PCBM) or indene-C_{60} bisadduct (ICBA)/bathocuproine (BCP)/Al, in which PEDOT:PSS acted as hole transport layer and PCBM or ICBA as electron transport layer, achieving a PCE of 3.9%.[35] Very promising PCE of over 15% has been demonstrated by optimizing perovskite morphology and the interface.[36-38] Several other inorganic counterparts with appropriate work functions have been tested as efficient alternatives to PEDOT:PSS, such as CuSCN,[39,40] NiO_x,[41-44] V_2O_5,[45] and so on. In particularly, NiO_x has shown the greatest promise due to its low cost, superior stability, high optical transmittance, and appropriate energy levels. The suitable work function of NiO_x facilitates hole transport and electron blocking as well as aligns well to the highest occupied molecular orbital (HOMO) level of MAPbI$_3$.[46,47] Seok and coworkers have reported a well-ordered nanostructured NiO$_x$ film prepared by pulsed laser deposition method.[47] The as-prepared NiO$_x$ film possessed good optical transparency and electrical conductivity, resulting in an overall PCE of 17.3% with a high FF of 0.81. Jen and coworkers used a simple and easy processable method, namely low-temperature combustion process, to prepare Cu-doped NiO$_x$ as hole transport layer for high-performance PHJ PSCs.[44]

Despite those impressive conversion efficiencies, the stability issue of PSCs has been considered as the most serious obstacle toward the commercialization. Several reports have indicated the instability of PSCs is mainly caused by the degradation of inorganic–organic hybrid lead perovskite materials and charge transport materials as well as the interface between them.[48–54] The inorganic–organic hybrid lead perovskite materials are sensitive to moisture, oxygen, ultraviolet (UV) light, thermal stresses, and some reactive dopants in charge transport layers. To achieve good reproducibility and long lifetime for PSCs with excellent efficiency, the degradation issues of perovskite and the device stability must be urgently addressed. This chapter summarizes the factors that influence the device stability. In addition, the strategies for stability improvement are proposed. This may give some insights for stability enhancement in PSCs, and lead to understand how far one is to the really stable devices. It is hoped that this chapter will be a useful report for the perovskite PV community.

II. Stability of Inorganic–Organic Hybrid Lead Perovskites and Their Solar Cells

High PCE and long-term stability are the two key requirements for commercialization of solar cells. At present, the relatively excellent performance has been achieved for the PSCs based on inorganic–organic hybrid lead perovskites, however, the device stability has still been a serious obstacle toward their practical application. The instability of perovskite devices mainly comes from the degradation of perovskite materials. Moisture, UV light, and temperature are susceptible to be other main factors that cause perovskite degradation.

A. Effect from Moisture

Among the influencing factors, moisture has been considered as the most challenging part. So far, several mechanisms have been reported on perovskite degradation caused by moisture. In 2014, Walsh and coworkers proposed a simple acid–base reaction mechanism for degradation of $MAPbI_3$ as shown in Figure 1.[48] They proposed that a single water molecule is sufficient to trigger the degradation process. H_2O firstly reacts with $MAPbI_3$ and forms the intermediate of $[(CH_3NH_3^+)_{n-1}(CH_3NH_2)_n PbI_3][H_3O]$. Then the intermediate decomposes into HI, CH_3NH_2, PbI_2, and H_2O. An excess of water is required to dissolve the HI and CH_3NH_2 by-products. Once the degradation process does happen, it will continuously go on until the H_2O is saturated with HI or the vapor pressure of CH_3NH_2 reaches equilibrium.

Wang and co-workers investigated the effect of ambient air and suggested the following degradation steps[49,55]:

$$CH_3NH_3PbI_3 \xleftrightarrow{H_2O} CH_3NH_3I(aq) + PbI_2(s) \qquad (1)$$

$$CH_3NH_3PbI(aq) \longleftrightarrow CH_3NH_2(aq) + HI(aq) \qquad (2)$$

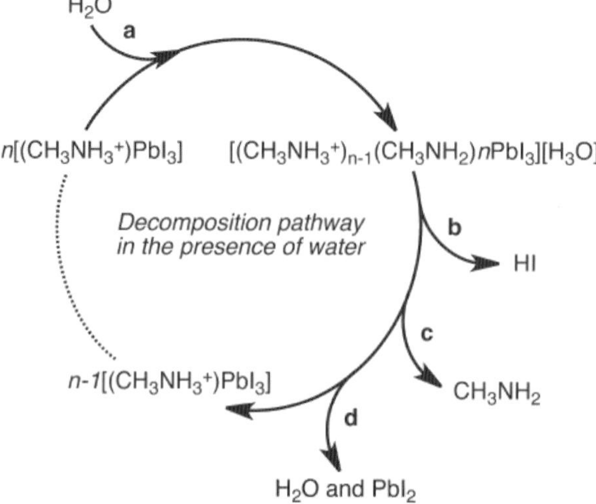

Figure 1. Possible decomposition pathway of hybrid halide perovskites in the presence of water. Taken from Ref. 48 with permission of the American Chemical Society.

$$4HI + O_2 \longleftrightarrow 2I_2(s) + 2H_2O \tag{3a}$$

$$2HI(aq) \xleftrightarrow{\text{hi}} H_2 \uparrow + I_2(s) \tag{3b}$$

They demonstrated that in the presence of H_2O, $MAPbI_3$ firstly decomposes into MAI solution and PbI_2, and then the MAI solution decomposes into CH_3NH_2 and HI. The resulted HI either reacts with oxygen in the air or takes on a photochemical reaction under light condition to generate I_2. X-ray diffraction (XRD) spectra of the $MAPbI_3$ film before and after degradation shown in Figure 2 assuredly verify degradation process as proposed earlier. All the original peaks of $MAPbI_3$ disappeared, and peaks of hexagonal 2H polytype PbI_2 and orthorhombic I_2 appeared.

Kamat and coworkers suggested another different degradation mechanism, which was confirmed by Kelly and coworkers laterly.[56,57] By characterization of the ground-state and excited-state optical absorption properties in combination with probing morphology and crystal structure of $MAPbI_3$, Kamat and coworkers proposed that $MAPbI_3$ should be able to complex with H_2O, forming a hydrate product similar to $MA_4PbI_6 \cdot 2H_2O$, rather than simply to revert to PbI_2 under H_2O exposure.[56] By comparing the UV–visible (UV–vis) absorption spectroscopy of $MAPbI_3$ film and the pure PbI_2 film stored in air with 90% relative humility at room temperature in the dark for 14 days, they found that the $MAPbI_3$ films were not converted to PbI_2 directly. The degradation of $MAPbI_3$ film does not give rise to the enhancement of absorption spectra at 500 nm, which is the characteristic peak of PbI_2. XRD patterns were recorded to further reveal the effects of humidity on the perovskite structure. The appearance of several new peaks in XRD patterns can be assigned to $MA_4PbI_6 \cdot 2H_2O$ crystals. Therefore, the degradation of $MAPbI_3$ in humid air was suggested to be described as:

$$CH_3NH_3PbI_3 + H_2O \longleftrightarrow (CH_3NH_3)_4 PbI_6 \cdot 2H_2O \tag{4}$$

Kelly and coworkers designed an experimental setup (shown in Figure 3a), which could precisely control the humility to quantitatively and systematically investigate the $MAPbI_3$ degradation processes.[57]

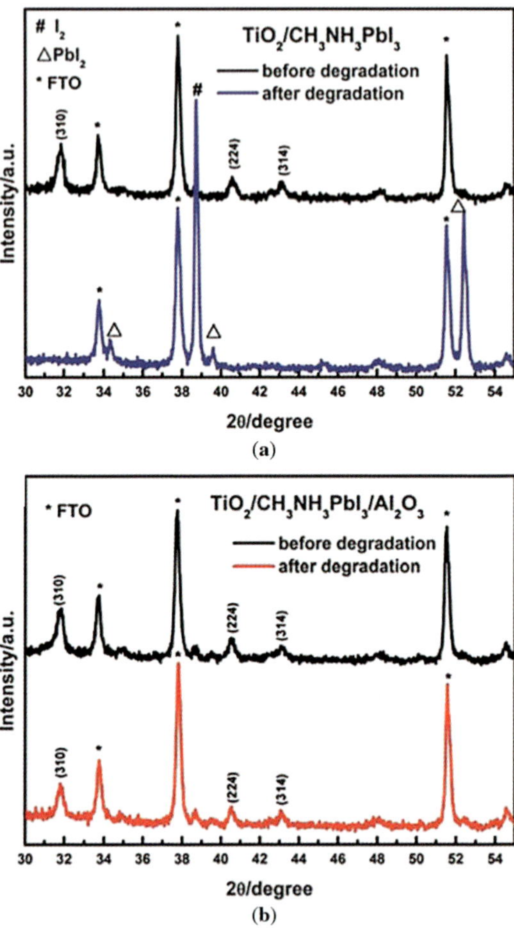

Figure 2. XRD patterns of sensitized films. (**a**) Film of TiO$_2$/CH$_3$NH$_3$PbI$_3$ before and after degradation. (**b**) Film of TiO$_2$/CH$_3$NH$_3$PbI$_3$/Al$_2$O$_3$ before and after degradation. The degradation time is 18 h. Taken from Ref. 49 with permission of the Royal Society of Chemistry.

In situ absorption spectroscopy was performed to monitor perovskite phase changes in degradation process (Figure 3b). Powder XRDs were also carried out on bulk sample of MAPbI$_3$. The results are shown in Figure 3c. The powder XRD patterns demonstrated that the water-added

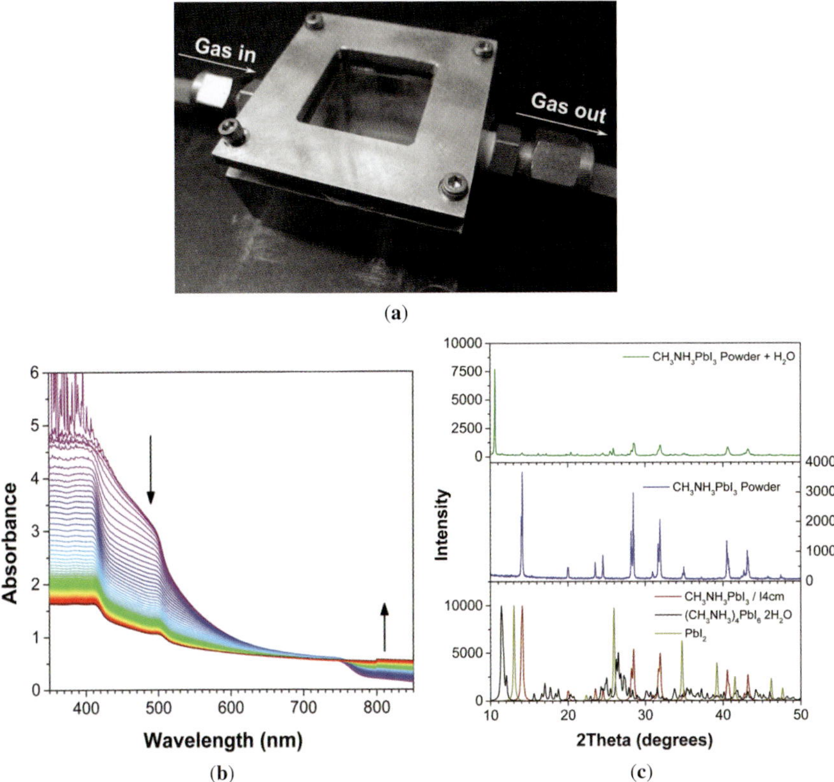

Figure 3.   (a) Sample holder for in situ UV–vis spectroscopy; (b) UV–vis spectra of a MAPbI$_3$ film exposed to N$_2$ gas with RH = 98 ± 2%, acquired at 15-min intervals. (c) Powder XRD patterns for MAPbI$_3$ powder, before and after the addition of the trace amount of water. The calculated powder patterns for tetragonal MAPbI$_3$, MA$_4$PbI$_6$·2H$_2$O, and PbI$_2$ are shown for comparison. Taken from Ref. 57 with permission of the American Chemical Society.

perovskite sample consists of a mixture of MA$_4$PbI$_6$·2H$_2$O and residual MAPbI$_3$. Thus, a hydrated intermediate containing isolated PbI$_6^{4-}$ octahedra was demonstrated as the first step of the perovskite degradation:

$$4CH_3NH_3PbI_3 + 2H_2O \longleftrightarrow \left(CH_3NH_3\right)_4 PbI_6 \cdot 2H_2O + 3PbI_2 \quad (5)$$

B. Effect from Temperature

Annealing is necessary for the formation of the perovskite crystal structure in a typical solution process.[36,58,59] In addition to this, temperature has great effect on the crystal structure and phase transition of the organic–inorganic hybrid perovskites.[60] Tolerance factor (t) has been proposed as an important index to evaluate the stability of AMX_3 perovskite. The tolerance factor can be described as:

$$ t = \frac{r_A + r_X}{\sqrt{2}\left(r_M + r_X\right)} \tag{6} $$

where r_A, r_M, and r_X are the ionic radii for the ions in the A, M, and X sites, respectively. Stable perovskites can be formed when the t values are between 0.8 and 1.[61,62] Therefore, in order to get a stable perovskite structure, the size of ionic radius should be carefully selected. In addition to the ionic radius, there are other aspects that significantly affect the perovskite phase transition and structural stability, such as temperature. As early as in 1987, Weber and coworkers investigated the relationship between temperature and the structure of methylammonium trihalide $MAPbX_3$ (X = Cl, Br, I).[62] Their results demonstrated that at room temperature $MAPbBr_3$ and $MAPbCl_3$ present a cubic structure. However, $MAPbI_3$ forms a tetragonal structure when the temperature increases to 327.4 K. The tetragonal phase of $MAPbI_3$ transforms to cubic phase along with a slight distortion of the PbI_6 octahedra around the c axes. Whereas, lowering the temperature would result in the transition from the tetragonal phase to orthorhombic phase. Baikie et $al.$ investigated the temperature-dependent phase transition for $MAPbX_3$ perovskite in detail and gave a description of the preparation, structural characterization, and physical characteristics of $MAPbI_3$.[11] In situ powder XRD and single crystal XRD characteristics were employed monitor the phase transition by varying temperature. The variation in lattice parameters with temperature indicates that $MAPbI_3$ undergoes a tetragonal to orthorhombic phase transition at approximately 161 K and a tetragonal to cubic phase transition at temperature of 329.15 K. This agrees well with the previous reports by Weber and coworkers.[62] The phase structure of $MAPbX_3$ perovskite strongly influences their electronic and optical properties.[15] For PSCs that are working under sun illumination, the temperature generated by long light illumination would cause phase transition and then affect the performance of solar cells. The PV metrics of $MAPbI_3$-based PSCs over a wide temperature range from 80 to 360 K have been investigated by Wang and coworkers.[63] They found that the PV metrics were strongly affected by the phase transition from the tetragonal to the orthorhombic phase. A maximum value of open-circuit voltage reaches at about 200 K, which is close to the phase transition from tetragonal to the orthorhombic phase. The photocurrent is remarkably stable down to 240 K but drops precipitously upon approaching and below the phase transition temperature at 160 K. Pisoni *et al.* reported the temperature-dependent thermal conductive properties of $MAPbI_3$.[64] Their research results demonstrated that for both large single crystals and the polycrystalline form, $MAPbI_3$ has a very low thermal conductivity. The ultralow thermal conductivity in $MAPbI_3$ is highly dependent to its particular crystal structure.

C. Effect from Photon

Due to the appropriate energy levels, the TiO_2 has become the most commonly used charge transfer material in PSCs, in the form of either compact TiO_2 or mesoporous TiO_2.[63,65–68] However, some experiments have indicated that when the TiO_2 was used as photoanode, the PSC devices' performance suffered from a rapid decay even though encapsulated in an inert atmosphere.[50,51,69,70] In 2013, Snaith and coworkers uncovered the reason for the instability of PSCs under illumination for the first time.[50] After excluding the influence of $MAPbI_3$ itself and hole transport layer spiro-OMeTAD carefully, they proposed that oxygen vacancies in the TiO_2 serve as deep electronic trap sites, which can bind with photo-induced electrons from perovskite and cause the degradation of device performance. The hypothesized degradation mechanism is shown in Figure 4. It is well-known that the TiO_2 particles contain oxygen vacancies, particularly at the surface.[71] Those oxygen vacancies are effectively deep electron-donating sites and can easily adsorb oxygen (Figure 4a). UV illumination could excite TiO_2 to form an electron-hole pair. The hole in the valence band of TiO_2 would recombine with the electron at the

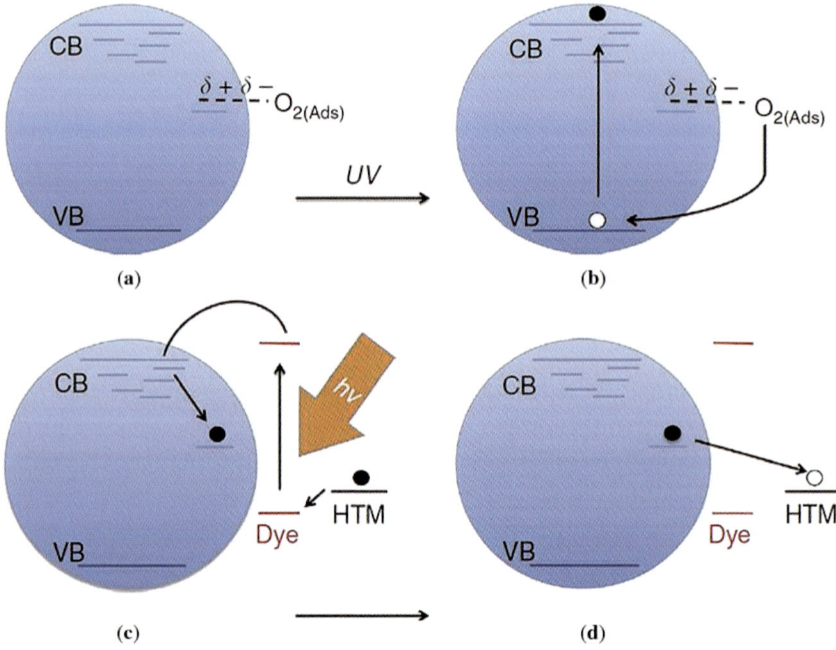

Figure 4. Degradation mechanism under UV illustration. Taken from Ref. 50 with permission of Nature Publishing Group.

oxygen adsorption site, resulting in desorption of oxygen (Figure 4b). In this case, a free electron in the conduction band of TiO_2 and a positively charged, unfilled oxygen vacancy site at the TiO_2 surface were left behind. The excess holes in heavily p-doped spiro-OMeTAD will readily recombine with the free electron and the photo-induced electrons from pigments will be trapped by the oxygen vacancy sites (Figure 4c). Finally, the holes on the HTM will recombine with the immobile trapped electrons (Figure 4d). The presence of oxygen could decrease the number of empty deep trap sites because they could adsorb on the TiO_2 to pacify these sites.

Another degradation mechanism for PSCs under light illumination was proposed by Nishino and coworkers.[51] They examined the light durability of $MAPbI_3$-based solar cells without encapsulation. The device architectures were (a) $FTO/TiO_2/MAPbI_3/CuSCN/Au$ and (b) $FTO/TiO_2/Sb_2S_3/MAPbI_3/CuSCN/Au$. As shown in Figure 5a, the conversion efficiency of devices with Sb_2S_3 was maintained 65% of the initial conversion efficiency without encapsulation after the 12-h light irradiation. However, the stability of the solar cell without Sb_2S_3 was very poor. The PCE decreased drastically within 5 h and was very close to zero within 12 h. Therefore, the PCE was significantly affected by the Sb_2S_3 layer in the TiO_2/$MAPbI_3$ interface. After a series of rigorous characterizations such as reflectance absorption spectra, XRD patterns, incident photon-to-current conversion efficiency (IPCE) spectra and Fourier transform infrared spectroscopy (FTIR) spectra, a reaction scheme used to explain the degradation effect of the $MAPbI_3$ layer against the light exposure was putted forward and shown in Figure 5b and c. By overnight light exposure, the $MAPbI_3$ layer without Sb_2S_3 was proposed to change to PbI_2, losing CH_3NH_2 and HI as shown in eq 4 and Figure 5b. On the other hand, the $MAPbI_3$ layer can be protected and made durable against light exposure with Sb_2S_3 insert layer (Figure 5c). Therefore, it is the interface between $MAPbI_3$ and TiO_2 that the decomposition of $MAPbI_3$ occurs at. As a typical n-type semiconductor, the TiO_2 is also a typical photocatalyst for environmental purification, such as reduction of CO_2, decomposition of organic compounds, and splitting of water.[72–76] The TiO_2 has a strong ability to extract electrons form electron-rich materials. Hence, electron extraction from iodide anion by the TiO_2 plays the driving force of the decomposition. Equations 7–10 could explain the possible reaction at the TiO_2 surface:

$$4CH_3NH_3PbI_3 \longleftrightarrow PbI_2 + CH_3NH_2 \uparrow + HI \uparrow \tag{7}$$

$$2I^- \longleftrightarrow I_2 + 2e^- \tag{8}$$

$$4CH_3NH_3^+ \longleftrightarrow 3CH_3NH_2 \uparrow + 3H^+ \tag{9}$$

$$I^- + I_2 + 3H^+ + 2e^- \longleftrightarrow 3HI \uparrow \tag{10}$$

First, the electron at I^- is extracted by the TiO_2. Consequently, the structure of $MAPbI_3$ is deconstructed, resulting in I_2 (Figure 5b and eq 5). Then, reaction 6 can be moved forward by a continuous elimination of H^+

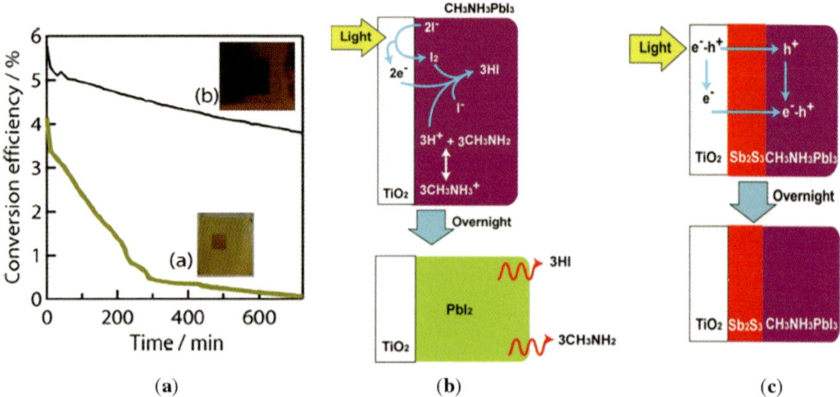

Figure 5. (a) Variation of photoenergy conversion efficiencies of solar cells during light exposure (AM1.5, 100 mW·cm^{-2}) without encapsulation in air for 12 h: (a) FTO/TiO$_2$/MAPbI$_3$/CuSCN/Au   and   (b)   FTO/TiO$_2$/Sb$_2$S$_3$/MAPbI$_3$/CuSCN/Au; Degradation scheme of MAPbI$_3$ PSCs during light exposure test: (**b**) without Sb$_2$S$_3$ layer and (**c**) with Sb$_2$S$_3$ layer. Taken from Ref. 51 with permission of the American Chemical Society.

through reaction 7 and evaporation of CH$_3$NH$_2$ (bp 17°C). Finally, I$_2$ is reduced by the extracted electrons at the interface between TiO$_2$ and MAPbI$_3$ (Figure 5b). Inserting a blocking layer of Sb$_2$S$_3$ between TiO$_2$ and perovskite MAPbI$_3$ can significantly enhance the device stability under light illumination.

D. Stability of Charge Transport Materials

Apart from effect of environmental factors and the degradation of inorganic–organic lead hybrid perovskite itself, charge transport materials also significantly influence the long-term stability of PSCs.[77–79] Charge transport layers both play an important role for transporting and blocking the charges. A wide number of HTMs ranging from classical semiconducting polymers to small molecules have been synthesized and investigated in combination with perovskite absorber.[24,80–83] Hitherto spiro-OMeTAD has been the mostly effective HTM in PSCs. However, the pristine spiro-OMeTAD suffers from low hole mobility and low conductivity due to the large intermolecular distances.[84–86] Usually, 4-tert-butylpyridine (TBP) and bis(trifluoromethane)sulfonimide lithium salt (Li-TFSI) are used to dope spiro-OMeTAD[49,87] in order to improve its electronic properties and polarity which increases the contact between $MAPbI_3$ and hole transport layer. However, the additive TBP is a polar solvent similar to g-butyrolactone and can dissolve the perovskite, further causing the instability of the PSCs. In addition, deliquescent behavior of Li-TFSI may tend to introduce moisture into the perovskite structure, thus aggravating its degradation. Li and coworkers studied the corrosion process of $MAPbI_3$ perovskite at the existence of TBP by UV–vis spectroscopy and X-ray photoelectron spectroscopy (XPS).[53] Upon the dropping of TBP, the perovskite films faded out under nitrogen atmosphere. Figure 6a shows the optical images of these films. The inter-reaction between $MAPbI_3$ and TBP was confirmed by UV–vis absorption spectroscopy. As shown in Figure 6a, the absorption was nearly close to zero after the spin coating of TBP.

It was found that PbI_2 reacts with TBP to form a transparent yellow liquid solution in room temperature. The solution was then dried under vacuum and a lighter yellow powder was obtained. By comparing the XPS spectra of the obtained yellow powder and PbI_2 powder, Li *et al.* found out that the binding energy of Pb 4f7/2 for TBP-treated powder shifted to a lower energy level from 139.02 to 138.47 eV as shown in Figure 6a and b. Those results indicate that a complex is formed by the interaction of TBP with the center lead metal, that is $[PbI_2 \times TBP]$.

On the other hand, PEDOT:PSS and PCBM are the most common used organic charge transport materials in inverted PHJ PSCs.[36,88,89] However, it was found that exposure in ambient air could cause the degradation of PCBM layer through adsorption of oxygen or water[52,90] and break the hydrogen bonds between individual PEDOT:PSS grains within the layer,[54,91] resulting in a significant loss in cohesion. As shown in Figure 7, the lowest occupied molecular orbital (LUMO) of PCBM increases after air exposure due to the water-PCBM interaction identified by the XPS spectra. The degradation of PCBM would result in a large contact resistance which decreases the device performance.

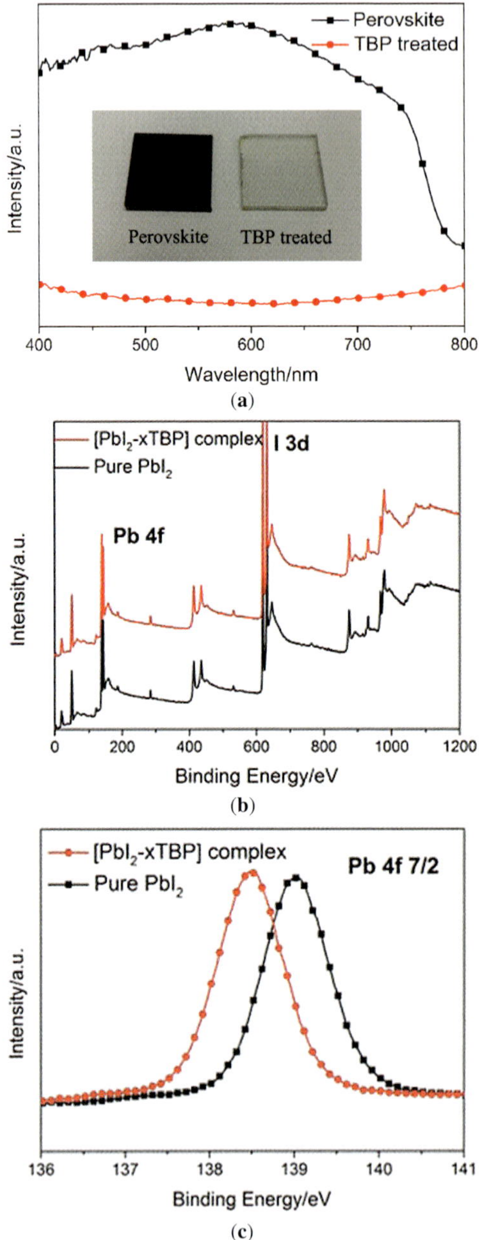

Figure 6. (a) UV–vis spectra of TiO$_2$/perovskite film and TBP-treated film, the photos of the films were inserted; (**b**) Overall XPS spectra; and (**c**) Pb 4f7/2 XPS spectra. Taken from Ref. 53 with permission of the Royal Society of Chemistry.

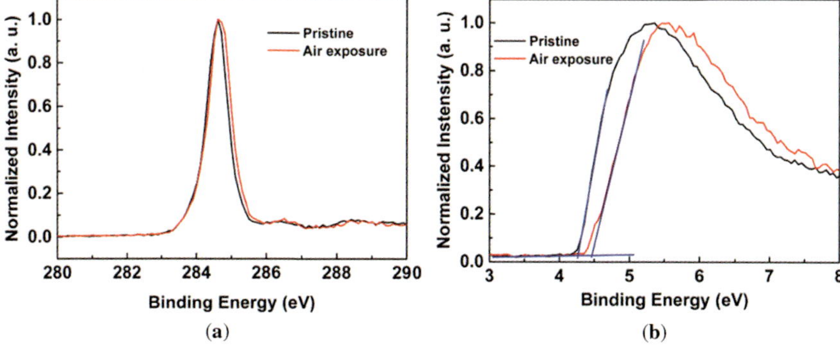

Figure 7.   (**a**) C 1s XPS of PCBM for with and without exposure to air (short exposure time during measurement) (**b**) UPS of PCBM with and without air exposure. Taken from Ref. 52 with permission of Nature Publishing Group.

III. Solution to Increase the Stability of Perovskite Solar Cells

A. Strategy for Reduction of Perovskite Degradation

One promising strategy to enhance the stability of perovskite is composition engineering to form perovskite materials with mixed cations or mixed halides. It was reported that iodide-chloride mixed halide perovskite $MAPbI_xCl_{3-x}$ was more stable than single-halide perovskite $MAPbI_3$ in ambient atmosphere.[23] Seok and coworkers also found that the stability of perovskite in humid air could be significantly improved by doping $MAPbI_3$ with bromide.[92] Besides, extensive researches have been done on formamidinium methylammonium lead iodide ($FAPbI_3$) due to its broad light absorption and good thermal stability.[12,93,94] The A cation in $FAPbI_3$ is larger than that in $MAPbI_3$, which would normally result in a higher tolerance factor t.[95] On the other hand, tuning the tolerance factor is also an effective way to stabilize perovskite structure for solar cell applications.[104] In 2015, Zhu and coworkers enhanced the stability of the perovskite structures by alloying $FAPbI_3$ with $CsPbI_3$.[95] The structure and stability of the mixed solid-state perovskite alloys $FA_{1-x}Cs_xPbI_3$ were studied by temperature-dependent XRD patterns and UV–vis spectroscopy. It was demonstrated that $FA_{1-x}Cs_xPbI_3$ alloy owned a higher stability

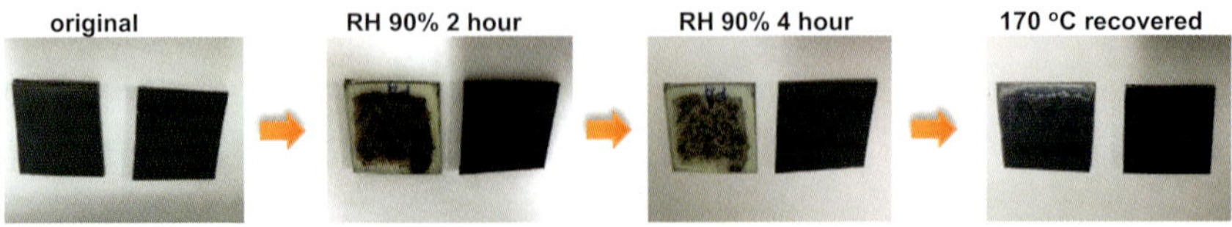

Figure 8. Photos of FAPbI$_3$ and FA$_{0.85}$Cs$_{0.15}$PbI$_3$ thin films under high-humidity conditions. Taken from Ref. 95 with permission of the American Chemical Society.

than FAPbI$_3$ under humidity conditions. As shown in Figure 8, no obvious color changes were observed for FA$_{0.85}$Cs$_{0.15}$PbI$_3$ films during aging; however, the FAPbI$_3$ film degraded from black to yellow just in 2 h under 90% humidity condition.

Inorganic perovskites that exhibit higher thermal stability than organic–inorganic hybrid counterparts have great potentialities to replace organic–inorganic hybrid perovskite to create stable solar cells. Two representative inorganic perovskites cesium lead trihalides (CsPbX$_3$) and cesium tin trihalides (CsSnX$_3$)[94,96–98] have already been applied as light absorbers in PVs. Snaith and coworkers proved that inorganic perovskite can work in solar cell devices by fabricating CsPbI$_3$-based planar-structured solar cells and a maximum 2.9% PCE was obtained.[96] Cahen and coworkers demonstrated that the organic cations was not necessary for high-performance PSCs by achieving a comparable PCE of CsPbBr$_3$-based solar cell to MAPbBr$_3$-based solar cells.[94,99] The thermal stability of CsPbBr$_3$ and MAPbBr$_3$ were compared by thermogravimetric analyses.[100] As shown in Figure 9, the first weight loss onset for CsPbBr$_3$ was about 580°C; however, the temperature was as low as 220°C for MAPbBr$_3$. It was approved that inorganic CsPbBr$_3$ is much more stable than the organic MAPbBr$_3$. The stability of solar cells based on MAPbBr$_3$ and CsPbBr$_3$ was compared in Figure 10. Devices fabricated with CsPbBr$_3$ demonstrated comparable PV performance to that of MAPbBr$_3$-based ones, but with much improved stability as shown in device aging studies. The work of inorganic PSCs paves the way for further developments, likely to lead to much more thermally stable solar cells and other optoelectronic devices.

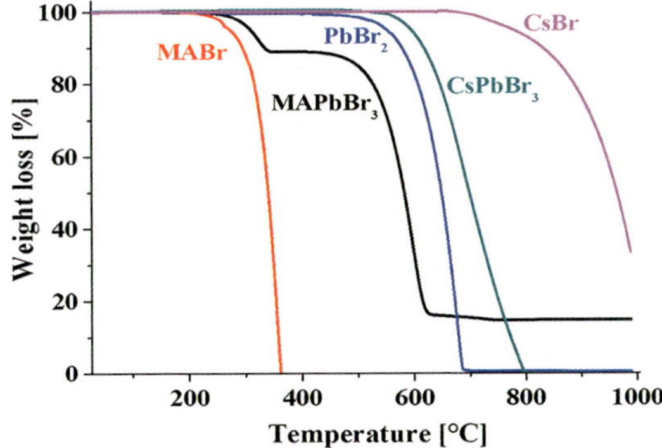

Figure 9.   Thermogravimetric analyses of methylammonium bromide (MABr), methylammonium lead bromide (MAPbBr$_3$), lead bromide (PbBr$_2$), cesium lead bromide (CsPbBr$_3$), and cesium bromide (CsBr). Taken from Ref. 100 with permission of the American Chemical Society.

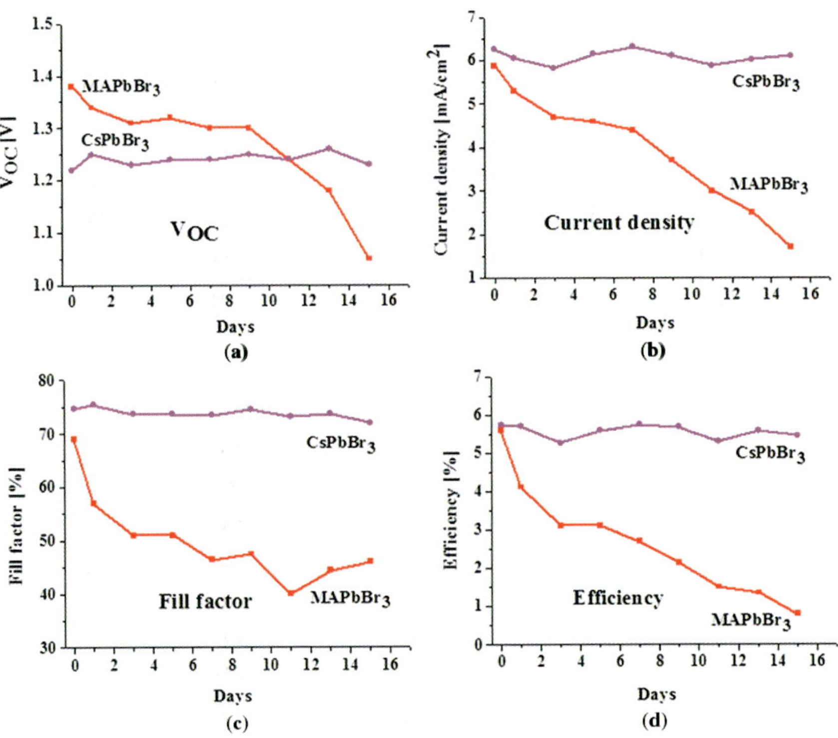

Figure 10.   Aging analysis of MAPbBr$_3$ and CsPbBr$_3$ cells. Figures show the cell parameters as a function of time. Taken from Ref. 100 with permission of the American Chemical Society.

B. Strategy for UV Light Degradation

As demonstrated earlier, under UV light illumination, the degradation of PSCs origins from the oxygen vacancies in the TiO_2 layer, which can trap photo-induced electrons in perovskite or extract electrons at I^-, thus deconstruct the structure of perovskite. Therefore, the effective routes to prevent UV degradation are to isolate the TiO_2 from UV light or perovskite layer, and even replace the n-type TiO_2 electron transport material with other efficient scaffold. As depicted in Figure 11a, Snaith and coworkers had adapted insulating mesoporous Al_2O_3 scaffold to replace TiO_2 and found significantly improved device stability.[50] Figure 11b shows the device performance parameters, which were measured under continuous simulated illumination of AM1.5, 76.5 $mW \cdot cm^{-1}$. It is quite evident that the Al_2O_3-based devices are considerably more stable than the TiO_2-based devices under UV light illumination. Contrary to the evolution of TiO_2-based cells under UV light exposure (Figure 11c), the Al_2O_3-based devices remained stable over the 1000-h exposure period and the photocurrent remains closed to 15 $mA \cdot cm^{-2}$.

Adding a UV filter in front of the TiO_2 is another route to mitigate UV degradation of PSCs. A transparent luminescent downshifting YVO_4:Eu^{3+} nano-phosphor layer was used as UV filter to prevent the TiO_2 from absorbing UV light (Figure 12).[101] Due to its specific electronic and magnetic properties, europium (Eu^{3+}) doped yttrium vanadate (YVO_4) nano-phosphor layer could absorb UV light up to 450 nm and emit long-wavelength light in the red region, which result in an enhancement in UV stability and improvement in PSCs' photocurrent.

In addition, isolation of the TiO_2 from perovskite is an effective strategy to enhance the stability of PSCs.[51] As shown in Figure 5b, inserting a surface blocking layer Sb_2S_3 at the interface between TiO_2 and $MAPbI_3$ could extend the distance between electrons and holes, preventing the recombination of holes at TiO_2 conduction band generated by UV irradiation and electrons in I^- anion of the $MAPbI_3$. Thus, the stability of the PSCs was enhanced during light exposure.

C. Strategies for Charge Transport Materials

Another choice to enhance the stability of PSCs is to find more stable charge transport materials. Han and coworkers have introduced an

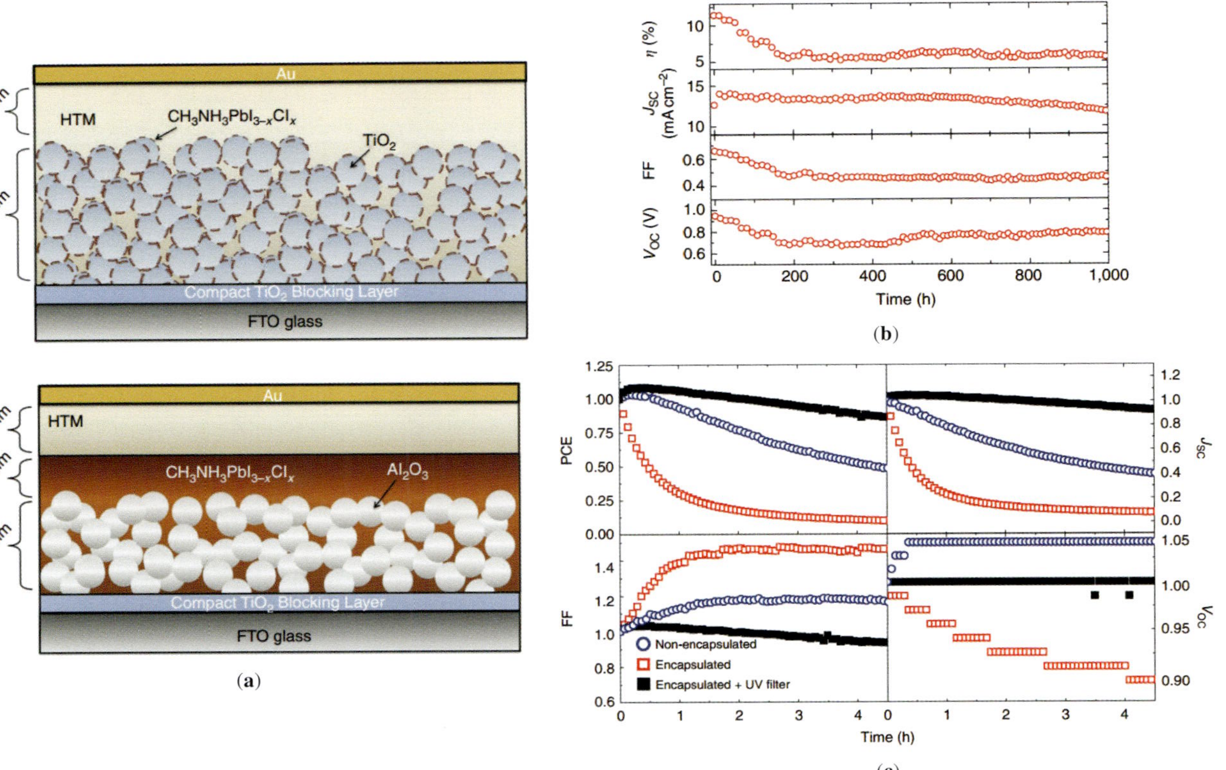

Figure 11. (a) Solar cell architectures; (b) Evolution of performance parameters of Al_2O_3-based PSC measured directly during aging under continuous illumination; (c) Evolution of normalized solar cell performance parameters for TiO_2-based solar cells. Taken from Ref. 50 with permission of Nature Publishing Group.

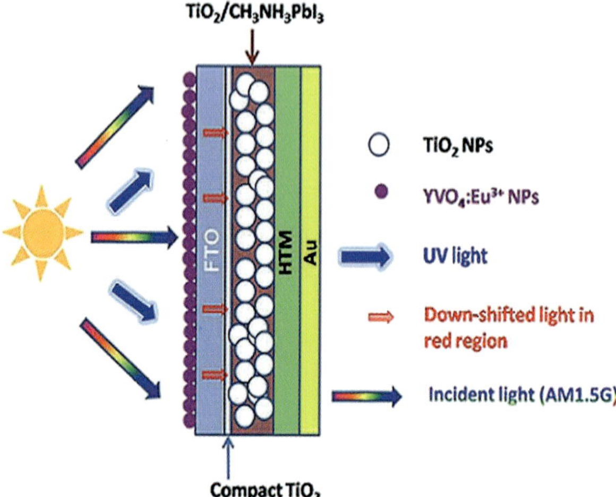

Figure 12. PSC structure with downshifting nano-phosphor layer spray coated on reverse of FTO glass. Absorbed UV light is downshifted to the red region. Taken from Ref. 101 with permission of AIP Publishing LLC.

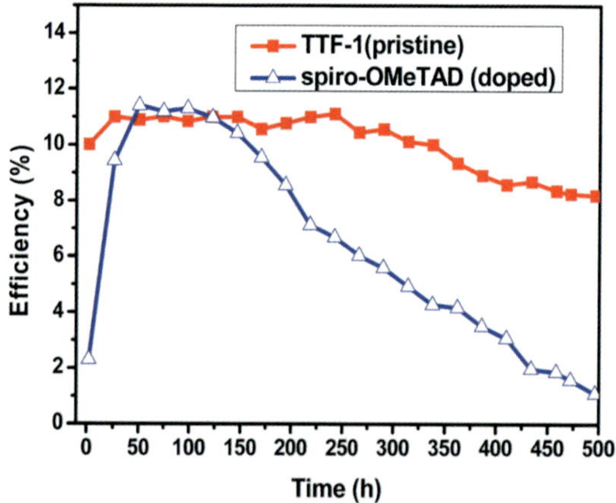

Figure 13. Efficiency variation of the optimized cells based on dopant-free TTF-1 (pristine) and p-type doped spiro-OMeTAD (doped). Taken from Ref. 79 with permission of the Royal Society of Chemistry.

efficient pristine hole transporting material, tetrathiafulvalene derivative (TTF-1), into PSCs without using any dopants.[79] As shown in Figure 13, the dopant-free TTF-1 based device exhibits more slight and slow efficiency degradation than the control device, which was based on doped

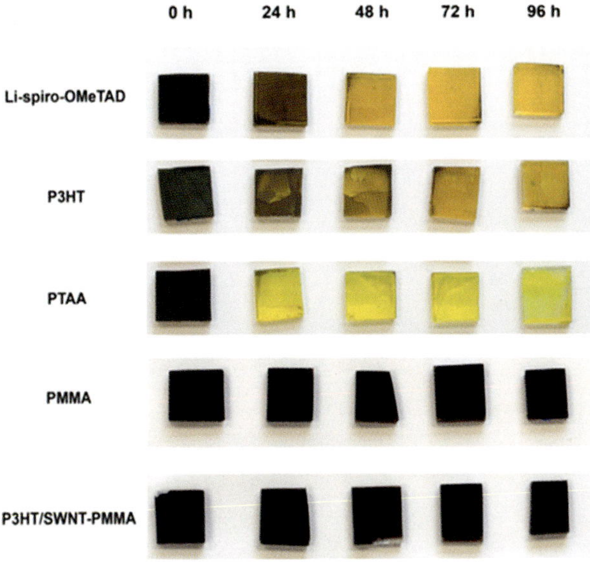

Figure 14.   Photos illustrating the visible degradation of the perovskite layers coated with different organic HTLs. Taken from Ref. 103 with permission of the American Chemical Society.

spiro-OMeTAD at a relative humidity of ≈40% in air without encapsulation. Other types of dopant-free HTMs have been reported showing promising application in PSCs as well. For example, Gong *et al.* proved that using dopant-free HTMs (DHPT-SC, DOPT-SC, and DEPTSC) can largely improve the cell stability and device performance.[102] Other HTMs that have protective effect on the perovskite structure could also retard the thermal degradation of PSCs by preventing $MAPbI_3$ from atmospheric moisture. Snaith and coworkers investigated the polymer-functionalized single-walled carbon nanotubes (SWNTs) embedded in an insulating polymer matrix as hole transport layer to replace the organic HTM.[103] As shown in Figure 14, the color of perovskite coated with commonly organic hole transporting materials spiro-OMeTAD, P3HT, and poly-(triarylamine) (PTAA), respectively, changed from dark brown to yellow quicker than that coated with SWNT-PMMA. The insulating polymer poly(methylmethacrylate) (PMMA) is nonhygroscopic and therefore could prevent the penetration of moisture into the perovskite structure while simultaneously enhance the device stability.

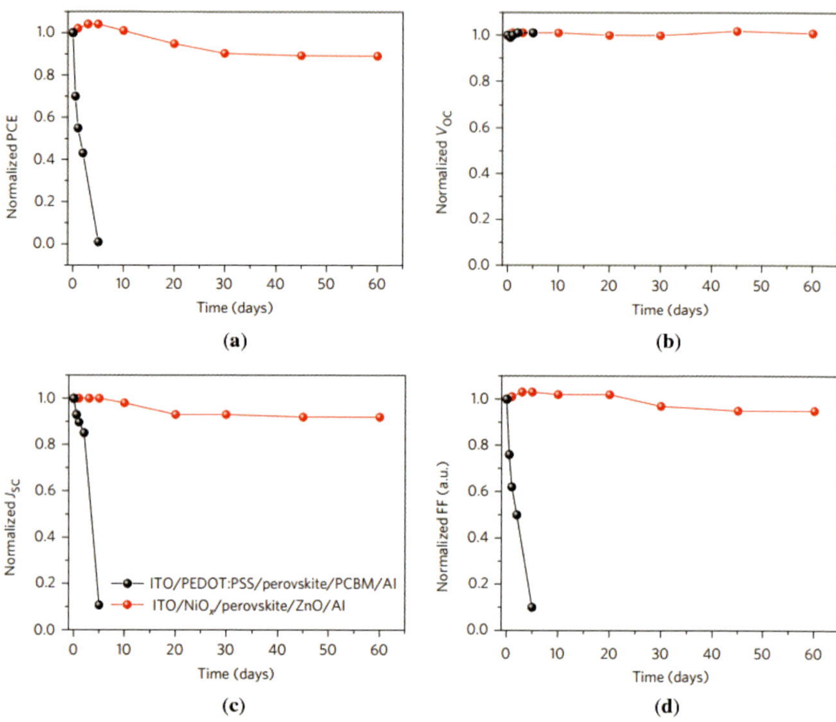

Figure 15. Device performances in an ambient environment without encapsulation of ITO/PEDOT:PSS/perovskite/PCBM/Al (black) and ITO/NiO$_x$/perovskite/ZnO/Al (red) structures. Taken from Ref. 52 with permission of Nature Publishing Group.

For the inverted PHJ devices, considering the sensitivity of PEDOT:PSS and PCBM to ambient moisture, stable inorganic metal oxides with high charge carrier mobility and matched energy level are used to increase the solar cell stability. Yang and coworkers have reported the p-type NiO$_x$ and n-type ZnO nanoparticles as hole and electron transport layers and the device stability against water and oxygen was significantly improved.[52] The stability of these devices using both inorganic and organic charge transport layers was monitored without encapsulation in an ambient environmental 25°C and with 30–50% humidity and the degradation of PV parameters were summarized in Figure 15a–d. The devices based on inorganic metal oxides NiO$_x$ and ZnO remain about

90% of their original efficiency after 60 days storage in air at room temperature, which are much more stable than the control devices made with organic transport layers PEDOT:PSS and PCBM. Those experimental results highlight the potential application of inorganic metal oxides in PSCs.

IV. Conclusions and Outlook

Due to their unique optical and electrical properties, perovskite materials have attracted much attention in the field of PV. Numerous research efforts have been taken to improve the conversion efficiency of PSCs and the highest conversion efficiency over 22.1% has been reached in 2016. By virtue of high efficiency, simple fabrication process, and low production cost, perovskite solar cells are very promising to commercialization. However, the long-term stability of PSC has always been an obstacle for its practical application. Moisture, UV light, temperature, and other components in the devices have important influence on device stability. Several degradation mechanisms of perovskite materials and device have been proposed, which provide fundamental understanding and give some insights for stability enhancement in PSCs. To overcome device stability issues and produce a stable PSC, various approaches have been explored, including adopting protecting layer, using alternative organic or inorganic charge transport materials, component engineering, interfacial engineering, and so on. Proper material processing and device handling technologies in PSCs will become more and more mature with deep understanding of perovskite materials degradation and device performance decay. Therefore, perovskite devices with high efficiency and long stability are expected to be realized for practical application in the near future.

V. References

1. The latest high efficiency was taken from NREL PV map: http://www.nrel.gov/ncpv/images/efficiency_chart.jpg

2. Kojima, A.; Teshima, K.; Shirai, Y.; Miyasaka, T. *J. Am. Chem. Soc.* **2009**, *131*, 6050–6051.

3. Eperon, G. E.; Burlakov, V. M.; Docampo, P.; Goriely, A.; Snaith, H. J. *Adv. Funct. Mater.* **2014**, *24*, 151–157.

4. Dong, Q. F.; Fang, Y. J.; Shao, Y. C.; Mulligan, P.; Qiu, J.; Cao, L.; Huang, J. S. *Science* **2015**, *347*, 967–970.

5. Nie, W. Y.; Tsai, H. H.; Asadpour, R.; Blancon, J. C.; Neukirch, A. J.; Gupta, G.; Crochet, J. J.; Chhowalla, M.; Tretiak, S.; Alam, M. A.; Wang, H. L.; Mohite, A. D. *Science* **2015**, *347*, 522–525.

6. Shi, D.; Adinolfi, V.; Comin, R.; Yuan, M. J.; Alarousu, E.; Buin, A.; Chen, Y.; Hoogland, S.; Rothenberger, A.; Katsiev, K.; Losovyj, Y.; Zhang, X.; Dowben, P. A.; Mohammed, O. F.; Sargent, E. H.; Bakr, O. M. *Science* **2015**, *347*, 519–522.

7. Lim, K. G.; Kim, H. B.; Jeong, J.; Kim, H.; Kim, J. Y; Lee, T. W. *Adv. Mater.* **2014**, *26*, 6461–6466.

8. Yu, H.; Wang, F.; Xie, F. Y.; Li, W. W.; Chen, J.; Zhao, N. *Adv. Funct. Mater.* **2014**, *24*, 7102–7108.

9. Xing, G. C.; Mathews, N.; Sun, S. Y.; Lim, S. S.; Lam, Y. M.; Grätzel, M.; Mhaisalkar, S.; Sum, T. C. *Science* **2013**, *342*, 344–347.

10. Sum, T. C.; Mathews, N. *Energy Environ. Sci.* **2014**, *7*, 2518–2534.

11. Baikie, T.; Fang, Y. N.; Kadro, J. M.; Schreyer, M.; Wei, F. X.; Mhaisalkar, S. G.; Graetzel, M.; White, T. J. *J. Mater. Chem. A* **2013**, *1*, 5628–5641.

12. Koh, T. M.; Fu, K. W.; Fang, Y. N.; Chen, S.; Sum, T. C.; Mathews, N.; Mhaisalkar, S. G.; Boix, P. P.; Baikie, T. *J. Phys. Chem. C* **2014**, *118*, 16458–16462.

13. Im, J. H.; Chung, J.; Kim, S. J.; Park, N. G. *Nanoscale Res. Lett.* **2012**, *7*, 353.

14. Kieslich, G.; Sun, S. J.; Cheetham, A. K. *Chem. Sci.* **2014**, *5*, 4712–4715.

15. Stoumpos, C. C.; Malliakas, C. D.; Kanatzidis, M. G. *Inorg. Chem.* **2013**, *52*, 9019–9038.

16. Hao, F.; Stoumpos, C. C.; Chang, R. P. H.; Kanatzidis, M. G. *J. Am. Chem. Soc.* **2014**, *136*, 8094–8099.

17. Noel, N. K.; Stranks, S. D.; Abate, A.; Wehrenfennig, C.; Guarnera, S.; Haghighirad, A. A.; Sadhanala, A.; Eperon, G. E.; Pathak, S. K.; Johnston, M. B.; Petrozza, A.; Herz, L. M.; Snaith, H. J. *Energy Environ. Sci,* **2014**, *7*, 3061–3068.

18. Hao, F.; Stoumpos, C. C.; Cao, D. H.; Chang, R. P. H.; Kanatzidis, M. G. *Nat. Photonics* **2014**, *8*, 489–494.

19. Cai, B.; Xing, Y. D.; Yang, Z.; Zhang, W. H.; Qiu, J. H. *Energy Environ. Sci.* **2013**, *6*, 1480–1485.

20. Edri, E.; Kirmayer, S.; Cahen, D.; Hodes, G. *J. Phys. Chem. Lett.* **2013**, *4*, 897–902.

21. Im, J. H.; Lee, C.R.; Lee, J. W.; Park, S. W.; Park, N. G. *Nanoscale* **2011**, *3*, 4088–93.

22. Kim, H. S.; Lee, C. R.; Im, J. H.; Lee, K. B.; Moehl, T.; Marchioro,A.; Moon, S. J.; Humphry-Baker, R.; Yum, J. H.; Moser, J. E.; Grätzel, M.; Park, N. G. *Sci. Rep.* **2012**, *2*, 591–597.

23. Lee, M. M.; Teuscher, J.; Miyasaka, T.; Murakami, T. N.; Snaith, H. J. *Science* **2012**, *338*, 643–647.

24. Heo, J. H.; Im, S. H.; Noh, J. H.; Mandal, T. N.; Lim, C. S.; Chang, J. A.; Lee, Y. H.; Kim, H. J.; Sarkar, A.; Nazeeruddin, M. K.; Grätzel, M.; Seok, S. *Nat. Photonics* **2013**, *7*, 486–491.

25. Ding, I. K.; Te´treault, N.; Brillet, J.; Hardin, B. E.; Smith, E. H.; Rosenthal, S. J.; Sauvage, F.; Grätzel, M.; McGehee, M. D. *Adv. Funct. Mater.* **2009**, *19*, 2431–2436.

26. Burschka, J.; Pellet, N.; Moon, S. J.; Humphry-Baker, R.; Gao, P.; Nazeeruddin, K.; Grätzel, M. *Nature* **2013**, *499*, 316–319.

27. Im, J. H.; Jang, I. H.; Pellet, N.; Grätzel, M.; Park, N. G. *Nat. Nanotechnol.* **2014**, *9*, 927–932.

28. Yang, W. S.; Noh, J. H.; Jeon, N. J.; Kim, Y. C.; Ryu, S.; Seo, J.; Seok, S. *Science* **2015**, *348*, 1234–1237.

29. Conings, B.; Baeten, L.; Dobbelaere, C. D,; Haen, J. D.; Manca, J.; Boyen, H. G. *Adv. Mater.* **2014**, *26*, 2041–2046.

30. Liu, M. Z.; Johnston, M. B.; Snaith, H. J. *Nature* **2013**, *501*, 395–398.

31. Zuo, L. J.; Gu, Z. W.; Ye, T.; Fu, W. F.; Wu, G.; Li, H. Y.; Chen, H. Z. *J. Am. Chem. Soc.* **2015**, *137*, 2674–2679.

32. Zhang, H.; Azimi, H.; Hou, Y.; Ameri, T.; Przybilla, T.; Spiecker, E.; Kraft, M.; Scherf, U.; Brabec, C. *J. Chem. Mater.* **2014**, *26*, 5190–5193.

33. Min, J.; Zhang, Z.G.; Hou, Y.; Quiroz, C. O. R.; Przybilla, T.; Bronnbauer, C.; Guo, F.; Forberich, K.; Azimi, H.; Ameri, T.; Spiecker, H.; Li, Y. F.; Brabec, C. *J. Chem. Mater.* **2015**, *27*, 227–234.

34. Zhou, H. P.; Chen, Q.; Li, G.; Luo, S.; Song, T. B.; Duan, H. S.; Hong, Z. R.; You, J. B.; Liu, Y. S.; Yang, Y. *Science* **2014**, *345*, 542–546.

35. Jeng, J. Y.; Chiang, Y. F.; Lee, M. H.; Peng, S. R.; Guo, T. F.; Chen, P.; Wen, T. C. *Adv. Mater.* **2013**, *25*, 3727–3732.

36. Wang, Q.; Shao, Y. C.; Dong, Q. F.; Xiao, Z. G.; Yuan, Y. B.; Huang, J. S. *Energy Environ. Sci.* **2014**, *7*, 2359–2365.

37. Xiao, Z. G.; Bi, C.; Shao, Y. C.; Dong, Q. Y.; Wang, Q.; Yuan, Y. B.; Wang, C. G.; Gao, Y. L.; Huang, J. S. *Energy Environ. Sci.* **2014**, *7*, 2619–2623.

38. Sun, C.; Wu, Z. H.; Yip, H. L.; Zhang, H.; Jiang, X. F.; Xue, Q. F.; Hu, Z. C.; Hu, Z. H.; Shen, Y.; Wang, M. K.; Huang, F.; CaO, Y. *Adv. Energy Mater.* **2016**, *6*, 1501534–1501544.

39. Ye, S. Y.; Sun, W. H.; Li, Y. L.; Yan, W. B.; Peng, H. T.; Bian, Z. Q.; Liu, Z. W.; Huang, C. H. *Nano Lett.* **2015**, *15*, 3723–3728.

40. Subbiah, A. S.; Halder, A.; Ghosh, S.; Mahuli, N.; Hodes, G.; Sarkar, S. K. *J. Phys. Chem. Lett.* **2014**, *5*, 1748–1753.

41. Cui, J.; Meng, F. P.; Zhang, H.; Cao, K.; Yuan, H. L.; Cheng, Y. B.; Huang, F.; Wang, M. K. *ACS Appl. Mater. Interfaces* **2014**, *6*, 22862–22870.

42. Jeng, J. Y.; Chen, K. C.; Chiang, T. Y.; Lin, P. Y.; Tsai, T. D.; Chang, Y. C.; Guo, T. F.; Chen, P.; Wen, T. C.; Hsu, Y. J. *Adv. Mater.* **2014**, *26*, 4107–4113.

43. Zhu, Z. L.; Bai, Y.; Zhang, T.; Liu, Z. K.; Long, X.; Wei, Z.H.; Wang, Z. L.; Zhang, L. X.; Wang, J. N.; Yan, F.; Yang, S. H. *Angew. Chem. Int. Ed.* **2014**, *53*, 12779–12783.

44. Kim, J. H.; Liang, P. W.; Williams, S. T.; Cho, N.; Chueh, C.C.; Glaz, M. S.; Ginger, D. S.; Jen, A. K. Y. *Adv. Mater.* **2015**, *27*, 695–701.

45. Docampo, P.; Ball, J. M.; Darwich, M.; Eperon, G. E.; Snaith, H. J. *Nat. Commun.* **2013**, *4*, 2761.

46. Yin, X. T.; Que, M. D.; Xing, Y. L.; Que, W. X. *J. Mater. Chem. A* **2015**, *3*, 24495–24503.

47. Park, J. H.; Seo, J.; Park, S.; Shin, S. S.; Kim, Y. C.; Jeon, N. J.; Shin, H. W.; Ahn, T. K.; Noh, J. H.; Yoon, S. C.; Hwang, C. S.; Seok, S. I. *Adv. Mater.* **2015**, *27*, 4013–4019.

48. Frost, J. M.; Butler, K. T.; Brivio, F.; Hendon, C. H.; Schilfgaarde, M.; Walsh, A. *Nano Lett.* **2014**, *14*, 2584–2509.

49. Niu, G. D.; Li, W. Z.; Meng, F. Q.; Wang, L. D.; Dong, H. P.; Qiu, Y. *J. Mater. Chem. A* **2014**, *2*, 705–710.

50. Leijtens, T.; Eperon, G. E.; Pathak, S.; Abate, A.; Lee, M. M.; Snaith, H. J. *Nat. Commun.* **2013**, *4*, 2885.

51. Ito, S.; Tanaka, S.; Manabe, K.; Nishino, H. *J. Phys. Chem. C* **2014**, *118*, 16995–17000.

52. You, J. B.; Meng, L.; Song, T. B.; Guo, T. F.; Yang, Y.; Chang, W. H.; Hong, Z. R.; Chen, H. J.; Zhou, H. P.; Chen, Q.; Liu, Y. S.; Marco, N. D.; Yang, Y. *Nat. Nanotechnol.* **2015**, *11*, 75–81.

53. Li, W. Z.; Dong, H. P.; Wang, L. D.; Li, N.; Guo, X. D.; Li, J. W.; Qiu, Y. *J. Mater. Chem. A* **2014**, *2*, 13587–13592.

54. Yip, H. L.; Jen, A. K. Y. *Energy Environ. Sci.* **2012**, *5*, 5994–6011.

55. Niu, G. D.; Guo, X. D.; Wang, L. D. *J. Mater. Chem. A* **2015**, *3*, 8970–8980.

56. Christians, J. A.; Herrera, P. A. M.; Kamat, P. V. *J. Am. Chem. Soc.* **2015**, *137*, 1530–1538.

57. Yang, J. L.; Siempelkamp, B. D.; Liu, D. Y.; Kelly, T. L. *ACS Nano* **2015**, *9*, 1955–1963.

58. Xiao, Z. G.; Dong, Q. F.; Bi, C.; Shao, Y. C.; Yuan, Y. B.; Huang, J. S. *Adv. Mater.* **2014**, *26*, 6503–6509.

59. Hsua, H. L.; Chen, C. P.; Chang, J. Y.; Yu, Y. Y.; Shen, Y. K. *Nanoscale* **2014**, *6*, 10281–10288.

60. Aharon, S.; Dymshits, A.; Rotem, A.; Etgar, L. *J. Mater. Chem. A* **2015**, *3*, 9171–9178.

61. Amat, A.; Mosconi, E.; Ronca, E.; Quarti, C.; Umari, P.; Nazeeruddin, M. K.; Grätzel, M.; Angelis, F. D. *Nano Lett.* **2014**, *14*, 3608–3616.

62. Poglitsch, A.; Weber, D. *J. Chem. Phys.* **1987**, *87*, 6373–6378.

63. Zhang, H.; Qiao, X. F.; Shen, Y.; Moehl, T.; Zakeeruddin, S. M.; Grätzel, M.; Wang, M. K. *J. Mater. Chem. A* **2015**, *3*, 11762–11767.

64. Pisoni, A.; Jaćimović, J.; Barišić, O. S.; Spina, M.; Gaál, R.; Forró, L.; Horváth, E. *J. Phys. Chem. Lett.* **2014**, *5*, 2488–2492.

65. Dharani, S.; Mulmudi, H. K.; Yantara, N.; Trang, P. T. T.; Park, N. G.; Graetzel, M.; Mhaisalkar, S.; Mathews, N.; Boix, P. P. *Nanoscale* **2014**, *6*, 1675–1679.

66. Kim, H. S.; Lee, J. W.; Yantara, N.; Boix, P. P.; Kulkarni, S. A.; Mhaisalkar, S.; Grätzel, M.; Park, N. G. *Nano Lett.* **2013**, *13*, 2412–2417.

67. Cao, K.; Zuo, Z. X.; Cui, J.; Shen, Y.; Moehl, T.; Zakeeruddin, S. M.; Grätzel, M.; Wang, M. K. *Nano Energy* **2015**, *17*, 171–179.

68. Cao, K.; Cui, J.; Zhang, H.; Li, H.; Song, J. K.; Shen, Y.; Cheng, Y. B.; Wang, M. K. *J. Mater. Chem. A* **2015**, *3*, 9116–9122.

69. Misra, R. K.; Aharon, S.; Li, B. L.; Mogilyansky, D.; Visoly-Fisher, I.; Etgar, L.; Katz, E. A. *J. Phys. Chem. Lett.* **2015**, *6*, 326–330.

70. Guarnera, S.; Abate, A.; Zhang, W.; Foster, J. M.; Richardson, G.; Petrozz, A.; Snaith, H. J. *J. Phys. Chem. Lett.* **2015**, *6*, 432–437.

71. Schwanitz, K.; Weiler, U.; Hunger, R.; Mayer, T.; Jaegermann, W. *J. Phys. Chem. C* **2007**, *111*, 849–854.

72. Zhang X. F.; Zhang B. Y.; Zuo Z. X.; Wang, M. K.; Shen, Y. *J. Mater. Chem. A* **2015**, *3*, 10020–10025.

73. Zhang X. F.; Zhang B. Y; Cao K, Brillet, J.; Chen, J. Y.; Wang, M. K.; Shen, Y. *J. Mater. Chem. A* **2015**, *3*, 21630–21636.

74. Habisreutinger, S. N.; Schmidt-Mende, L.; Stolarczyk, J. K. *Angew. Chem. Int. Ed.* **2013**, *52*, 7372–7408.

75. Yu, J. G.; Low, J.; Xiao, W.; Zhou, P.; Jaroniec, M. *J. Am. Chem. Soc.* **2014**, *136*, 8839–8842.

76. Matsumoto, K.; Makino, T.; Ebara, T.; Mizuguchi, J. *J. Chem. Eng. Japan* **2008**, *41*, 51–56.

77. Christians, J. A.; Fung, R. C. M.; Kamat, P. V. *J. Am. Chem. Soc.* **2013**, *136*, 758–764.

78. Zhang, L. Q.; Zhang, X. W.; Yin, Z. G.; Jiang, Q.; Liu, X.; Meng, J. H.; Zhao, Y. J.; Wang, H. L. *J. Mater. Chem. A* **2015**, *3*, 12133–1238.

79. Liu, J.; Wu, Y. Z.; Qin, C. J.; Yang, X. D.; Yasuda, T.; Islam, A.; Zhang, K.; Peng, W. Q.; Chen, W.; Han, L. Y. *Energy Environ. Sci.* **2014**, *7*, 2963–2967.

80. Ma, S. Y.; Zhang, H.; Zhao, N.; Cheng, Y. B.; Wang, M. K.; Shen, Y.; Tu, G. L. *J. Mater. Chem. A* **2015**, *3*, 12139–12144.

81. Li, H. R.; Fu, K.; Hagfeldt, A.; Grätzel, M.; Mhaisalkar, S. G.; Grimsdale, A. C. *Angew. Chem. Int. Ed.* **2014**, *53*, 4085–4088.

82. Xu, B.; Sheibani, E.; Liu, P.; Zhang, J. B.; Tian, H. N.; Vlachopoulos, N.; Boschloo, G.; Kloo, L.; Hagfeldt, A.; Sun, L. C. *Adv. Mater.* **2014**, *26*, 6629–6634.

83. Qin, P.; Tanaka, S.; Ito, S.; Tetreault, N.; Manabe, K.; Nishino, H.; Nazeeruddin, M. K.; Grätzel, M. *Nat. Commun.* **2014**, *5*, 3834–3836.

84. Leijtens, T.; Lim, J.; Teuscher, J.; Park, T.; Snaith, H. J. *Adv. Mater.* **2013**, *25*, 3227–3233.

85. Cai, B.; Xing, Y.; Yang, Z.; Zhang W. H.; Qiu, J. *Energy Environ. Sci.* **2013**, *6*, 1480–1485.

86. Burschka, J.; Dualeh, A.; Kessler, F.; Baranoff, E.; Cevey-Ha, N. L.; Yi, C. Y.; Nazeeruddin, M. K.; Grätzel, M. *J. Am. Chem. Soc.* **2011**, *133*, 18042–18045.

87. Abate, A.; Leijtens, T.; Pathak, S.; Teuscher, J.; Avolio, R.; Errico, M. E.; Kirkpatrik, J.; Ball, J. M.; Docampo, P.; Phersonc, I. M.; Snaith, H. J. *Phys. Chem. Chem. Phys.* **2013**, *15*, 2572–2579.

88. Xue, Q. F.; Hu, Z. C.; Liu, J.; Lin, J. H.; Sun, C.; Chen, Z. M.; Duan, C. H.; Wang, J.; Liao, C.; Lau, W. M.; Huang, F.; Yip, H. L.; Cao, Y. *J. Mater. Chem. A* **2014**, *2*, 19598–19603.

89. Choi, H.; Mai, C. K.; Kim, H. B.; Jeong, J.; Song, S.; Bazan, G. C.; Kim, J. Y.; Heeger, A. J. *Nat. Commun.* **2015**, *6*, 7348.

90. Bao, Q.; Liu, X.; Braun, S.; Fahlman, M. *Adv. Energy Mater.* **2014**, *4*, 1301272.

91. Dupont, S. R.; Novoa, F.; Voroshazi, E.; Dauskardt, R. H. *Adv. Funct. Mater.* **2014**, *24*, 1325–1332.

92. Noh, J. H.; Im, S. H.; Heo, J. H.; Mandal, T. N.; Seok, S. I. *Nano Lett.* **2013**, *13*, 1764–1769.

93. Eperon, G. E.; Stranks, S. D.; Menelaou, C.; Johnston, M. B.; Herz, L. M.; Snaith, H. J. *Energy Environ. Sci.* **2014**, *7*, 982–988.

94. Binek, A.; Hanusch, F. C.;Docampo, P.; Bein, T. *J. Phys. Chem. Lett.* **2015**, *6*, 1249–1253.

95. Li, Z.; Yang, M.; Park, J. S.; Wei, S. H.; Berry, J. J.; Zhu, K. *Chem. Mater.* **2016**, *28*, 284–292.

96. Eperon, G. E.; Paternò, G. M.; Sutton, R. J.; Zampetti, A.; Haghighirad, A. A.; Caciallib, F.; Snaith, H. J. *J. Mater. Chem. A* **2015**, *3*, 19688–19695.

97. Chen, Z.; Wang, J. J.; Ren, Y.; Yu, C.; Shum, K. *Appl. Phys. Lett.* **2012**, *101*, 093901.

98. Kumar, M. H.; Dharani, S.; Leong, W. L.; Boix, P. P.; Prabhakar, R. R.; Baikie, T.; Shi, C.; Ding, H.; Ramesh, R.; Asta, M.; Graetzel, M.; Mhaisalkar, S. G.; Mathews, N. *Adv. Mater.* **2014**, *26*, 7122–7127.

99. Kulbak, M.; Cahen, D.; Hodes, G. *J. Phys. Chem. Lett.* **2015**, *6*, 2452–2456.

100. Kulbak, M.; Gupta, S.; Kedem, N.; Levine, I.; Bendikov, T.; Hodes, G.; Cahen, D. *J. Phys. Chem. Lett.* **2016**, *7*, 167–172.

101. Chander, N.; Khan, A. F.; Chandrasekhar, P. S.; Thouti, E.; Swami, S. K.; Dutta, V.; Komarala, V. K. *Appl. Phys. Lett.* **2014**, *105*, 033904.

102. Gong, G. F.; Zhao, N.; Ni, D. B.; Chen, J. Y.; Shen, Y.; Wang, M. K.; Tu, G. L. *J. Mater. Chem. A* **2016**, *4*, 3661–3666.

103. Habisreutinger, S. N.; Leijtens, T.; Eperon, G. E.; Stranks, S. D.; Nicholas, R. J.; Snaith, H. J. *Nano Lett.* **2014**, *14*, 5561–5568.

104. Saliba, M.; Matsui, T.; Seo, J. Y.; Domanski, K.; Correa-Baena, J. P.; Nazeeruddin, M. K.; Zakeeruddin, S. M.; Tress, W.; Abate, A.; Hagfeldt, A.; Gratzel, M. *Energy Environ. Sci.* **2016**, *9*, 1989–1997.

7 Time-Resolved Photoconductivity Measurements on Organometal Halide Perovskites

Eline M. Hutter*, Tom J. Savenije*
and Carlito S. Ponseca Jr.[†,‡]

*Department of Chemical Engineering, Delft University
of Technology, van der Maasweg 9, 2629 HZ Delft, The Netherlands
[†]Division of Chemical Physics, Lund University,
Box 124, 221 00 Lund, Sweden
[‡]carlito.ponseca@liu.se

Table of Contents

List of Abbreviations

α	the absorption coefficient
$\Delta\sigma$	the photoconductivity
CB	conduction band
DC	direct current
F_A	the fraction of absorbed light
FAPbI$_3$	formamidinium lead triiodide
F_R	the fraction of reflected light
F_{R+T}	the total fraction of reflected and transmitted light
F_T	transmitted light
HOMO	the highest occupied molecular orbital
I_0	the incident intensity
I_R	the reflected light
IRF	instrumental response function
I_T	the transmitted light intensity
I_x	the light intensity in the sample at distance X
LUMO	the lowest unoccupied molecular orbital
MAPbI$_3$	methyl ammonium lead triiodide
MA$^+$	methyl ammonium cation
N_t	trap states
OMHP	organometal halide perovskite
p_0	background carriers
PCE	power conversion efficiency

PL	photoluminescence
SEM	scanning electron microscope
TA	transient absorption
THz	time-resolved terahertz
TRMC	microwave conductivity
TRPL	time-resolved photoluminescence
VB	valence band
XRD	the X-ray diffraction

I. Introduction

The meager performance of metal halide perovskite (methyl ammonium lead triiodide, $MAPbI_3$) as sensitizer in a dye-sensitized solar cell in 2009[1] caught the solar cell community by surprise, because its power conversion efficiency (PCE) uninterruptedly increased, reaching 22.1% in 2016.[2] This is in strong contrast to other photovoltaic technologies, for example, dye-sensitized and organic solar cells, where decades of meticulous research did not lead to these high performances. Such extraordinary development on perovskite-based materials has changed the landscape of solar cell research prompting many groups to refocus their efforts in this field, causing a snowball effect in its development. Recently, it has been shown that organometal halide perovskites can also be used in light-emitting diodes,[3] laser,[4] and even as catalyst for water photolysis.[5] Admittedly, there are many reports on the excited state dynamics of these materials on time scales of sub-picosecond to sub-second using diverse spectroscopic techniques. However, the behavior of charge carriers has remained vague to say the least while some of the results are rather inconsistent with each other. For example, the nature of initial photoproduct remains debated in the literature as to whether they are exciton or mobile charge carriers. This chapter presents an unambiguous approach in monitoring the charge carrier dynamics. By directly accessing the changes in the conductivity of the material on optical excitation, it is presented (1) how charged species are generated, (2) how mobile they are, (3) how these are injected into charge-specific electrodes, and (4) how they recombine. Finally, it is shown how transient absorption and photoluminescence (PL) can be utilized to complement these techniques.

This chapter is organized as follows. The first section presents the details of the time-resolved terahertz (THz) and time-resolved microwave conductivity (TRMC) techniques and introduces a kinetic model to analyze the data. The second section discusses the picosecond and nanosecond charge behavior of intrinsic perovskite material, that is, thin polycrystalline films of $MAPbI_3$ and $FAPbI_3$, as well as perovskites attached to metal oxides or organic electrodes. The third section disentangles the processes involved in the recombination decay dynamics in the nanosecond to microsecond time scales using TRMC measurements both in thin films and in single crystals.

II. Methods

A. Light Absorption

The absorption of light upon illumination by a short laser pulse is a key element in THz spectroscopy and TRMC measurements. The penetration of light through a material varies with the photon energy and as a result, the charge carrier generation profile is highly dependent on the excitation wavelength. In general, the transmitted light intensity (I_T) with respect to the incident intensity (I_0) is related to the absorption coefficient α:

$$\frac{I_T}{I_0} = e^{-\alpha L} \tag{1}$$

where L is the sample thickness. Note that this equation neglects reflection. However, because organometal halide perovskites (OMHPs) are highly reflective, I_0 has to be corrected for the reflected light (I_R). The light intensity in the sample at distance x (I_x) from the surface can be calculated from:

$$\frac{I_x}{I_0 - I_R} = e^{-\alpha x} \tag{2}$$

Figure 1 shows the absorption coefficient, α (cm^{-1}) of an $MAPbI_3$, calculated from absorption and transmission spectra of thin (50 and 300 nm) films. These spectra were recorded with a spectrophotometer

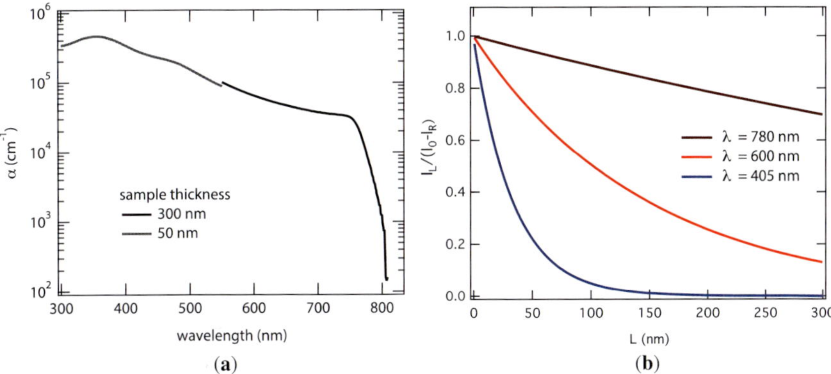

Figure 1. (a) Semi-logarithmic plot of the absorption coefficient of $MAPI_3$. (b) Initial charge carrier generation profile for different excitation wavelengths using the absorption coefficients from a: $1.2 \times 10^4 \cdot cm^{-1}$ for 780 nm, $6.8 \times 10^4 \cdot cm^{-1}$ for 600 nm, and $3 \times 10^4 \cdot cm^{-1}$ for 405 nm. As detailed in eq 2, the incident light is corrected for reflection.

equipped with an integrating sphere: the thin films were placed in front of the sphere to measure the fraction of transmitted light (F_T) and inside the sphere to detect the total fraction of reflected and transmitted light (F_{R+T}). The latter was used to calculate the fraction of absorbed light (F_A):

$$F_A = 1 - F_{R+T} \tag{3}$$

The fraction of reflected light (F_R) is determined by:

$$F_R = 1 - F_A - F_T \tag{4}$$

Analogous to eq 2, α is obtained from:

$$\frac{F_T}{1-F_R} = e^{-\alpha L} \tag{5}$$

For thick ($L > 300$ nm) perovskite films, the F_T at high excitation energies ($\lambda < 500$ nm) is too small to accurately measure and therefore eq 5 cannot be used to determine α. On the other hand, the light absorption of thin samples is not sufficient to resolve α close to the absorption onset. Therefore, films with different thicknesses were used to determine α.

Figure 1a shows that α is three orders of magnitude larger for $\lambda = 400$ nm than $\lambda = 800$ nm. Recalling eq 2, this means that varying the excitation wavelength of the laser used in spectroscopy measurements changes the initial concentration profile of the charge carriers. This is plotted in Figure 1b, showing the penetration of light through $MAPbI_3$ for excitation wavelengths of 405, 600, and 780 nm. From this, it can be deduced that more than 95% of 405 nm light is absorbed within 100 nm $MAPbI_3$. In other words, upon illuminating $MAPbI_3$ at 405 nm, the majority of photoexcited charges is initially located within 100 nm of the surface, independent of the total sample thickness. On the other hand, higher excitation wavelengths initially result in a lower concentration, but more homogeneous distribution of charge carriers.

B. Time-Resolved THz Spectroscopy

Pulsed femtosecond (80 fs) laser light with energy higher than the band gap of a perovskite material is used to generate charges. This leads to a difference in the conductivity of the material from its ground state value to its photoexcited value. Such change is the main result of the experiment and by changing the delay time between the pump (laser excitation) and the probe (THz), the evolution of the photoconductivity ($\Delta\sigma$) can be monitored from sub-picoseconds (instrument response) up to a few nanoseconds. Note, however, that it is possible that upon photoexcitation, tightly bound excitons, whose charge is neutral can also be generated. But, because an exciton does not have a net charge it will not lead to a change in photoconductivity and therefore is not detected in this specific experimental configuration. Likewise, electrons and holes optically generated but with a very low mobility (<0.005 cm^2/Vs), although should contribute to change the photoconductivity, may be masked by the contribution of other highly mobile charges and therefore cannot be discerned by the instrument. The transient photoconductivity, $\Delta\sigma$, can be calculated using the following equation:

$$\frac{\Delta\sigma}{F_A I_0 e} = \varphi \cdot \left(\mu_e + \mu_h\right) = -\frac{\Delta E_{exc}(\omega)}{E_{gs}(\omega)} \cdot \frac{\varepsilon_0 c}{F_A I_0 e} \tag{6}$$

where φ is photon-to-charge conversion ratio or quantum yield, μ_e and μ_h are the mobility of electrons and holes mobility, respectively, ΔE_{exc} is the change in the transmitted THz electric field after photoexcitation while E_{gs} are the transmitted THz electric field without excitation, ε_0 is permittivity of vacuum, c is velocity of light, F_A is the fraction of light absorbed at the excitation wavelength, I_0 is the fluence in photons/cm^2, and e is the elementary charge. The quantity that can be obtained from this $\varphi(\mu_e + \mu_h)$ has the units of mobility in cm^2/Vs.

The interplay between the time-dependent change in charge population and mobility defines the shape of the THz transient photoconductivity. On one hand, the rise in the photoconductivity elucidates generation of charged, mobile species and/or increase in mobility of the charges (further separation of charge carriers). On the other hand, decay represents decrease of the mobility (maybe due to relaxation) and/or disappearance of charge carriers either by recombination, or by injection into a low mobility acceptor material. The transient THz photoconductivity kinetics are collected by fixing the gating (delay line 1 of Figure 2) at the peak of the THz electric field and scanning the pump-probe delay (delay line 2 of Figure 2) within a desired time interval, typically up to few nanoseconds. It should be noted that at the earliest time scale, φ is assumed to be 1, while at longer times this quantity represents the fraction of the initial charge population at a particular time. If all absorbed photons are converted to charged species, φ equals unity. If an accurate determination of φ is not possible, a φ of 1 can be used to obtain a lower estimate of the mobility. To obtain the photoconductivity spectra, one should first collect the transmitted THz electric fields ΔE_{exc} and E_{gs} at a desired pump-probe delay time. This can be done by moving delay line 1 to map the THz electric field emitted by ZnTe crystal. The ratio of the Fourier transform of these THz electric fields will yield the amplitude and phase of the transient transmittance, and by using eq 6 can then be converted to real and imaginary parts of photoconductivity, respectively. The shape of the resulting photoconductivity spectra could be used to determine the mode of transport of charges. When charges are confined or restricted in distances smaller than the diffusion length of the material, which may be due to some traps, defects, or it hits a boundary where it backscatters, the real part of photoconductivity is positive while the imaginary part is negative.

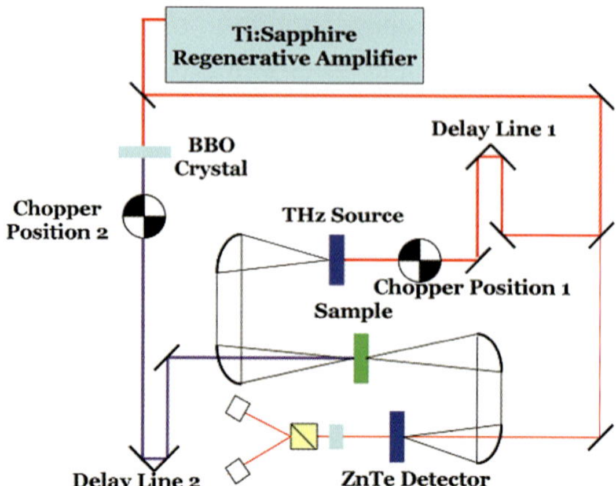

Figure 2. Schematic diagram of the time-resolved THz setup used in probing charge carrier dynamics in OMHP thin films. The pulsed THz radiation is generated and detected using ZnTe crystal while the pump laser pulse is generated by BBO Crystal. Photoconductivity kinetics is obtained by positioning delay line 1 at the peak of THz electric field and then scanning delay line 2 up to 1 ns. To obtain the photoconductivity spectra, delay line 2 is positioned at a desired pump-probe time and delay line 1 is scanned to map the THz electric fields, ΔE_{exc} and E_{gs}. Adapted from Ref. 7.

For long range, electron gas-like transport, both the real and imaginary parts are positive.[6]

C. Time-Resolved Microwave Conductivity

Analogous to THz spectroscopy, the TRMC technique can be used to study the dynamics of photoinduced charge carriers in low conductive semiconductor materials.[8–11] This technique is based on the interaction between the electric field component of microwaves (GHz regime) and mobile carriers. Pulsed laser light is used to excite electrons from the valence band (VB) to the conduction band (CB) of the semiconductor of interest. TRMC measurements have been extensively used to study charge carrier dynamics in organic systems, for example, conjugated polymers,[8,12] in which electrons are excited from the highest occupied molecular orbital (HOMO) to the lowest unoccupied molecular orbital

(LUMO). If photo-excitation of a material results in the generation of free, mobile charge carriers, the conductivity of the material is enhanced. By definition, electrical conductivity σ scales with the concentration of free charges n and their mobility μ according to:

$$\sigma = ne\Sigma\mu \tag{7}$$

With the TRMC technique, the change in conductivity between dark and after illumination, that is, $\Delta\sigma$, is studied. From here, the photoconductance ΔG can be determined, which will be described in more detail later in this chapter. Although an inhomogeneous excitation profile can result in a gradient in the conductivity throughout the sample, with TRMC the integrated change in conductivity over the film thickness is determined. Therefore, ΔG is proportional to the product of the total number of photoinduced free charges and their mobility. Absorption of microwaves by photoinduced charges reduces the microwave power P on the detector. This is schematically depicted in Figure 3. The normalized reduction in microwave power P ($\Delta P/P$) is related to ΔG by:

$$\frac{\Delta P}{P} = -K\Delta G \tag{8}$$

where K is a sensitivity factor. Note, that ΔP is negative while ΔG is positive. ΔP is recorded as function of time after the laser pulse and thus, the TRMC technique can be used to determine both the mobility and time-dependent concentration (i.e., the lifetime) of photoinduced free charges. Considering that both electrons and holes contribute to the photoconductance, ΔG is proportional to the sum of both their concentrations and mobilities (see Scheme 1). This is similar to THz photoconductivity (eq 6), but in contrast with frequently used time-resolved photoluminescence measurements, which specifically detects radiative recombination events. Therefore, the decays obtained with TRMC cannot be directly compared to PL transients, and careful data analysis is required. Additionally, PL does not necessarily originate from recombination between free mobile charges. Therefore, the processes observed with TRMC can be different from those detected with PL.

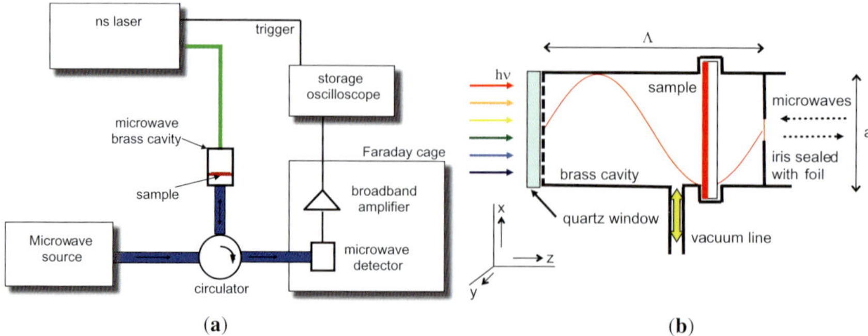

Figure 3. (**a**) schematic representation of the TRMC setup. Monochromatic microwaves are generated using a voltage-controlled oscillator (microwave source). The sample of interest is placed in a fully reflective microwave cell. Taken from Ref. 10 with permission of American Chemical Society. (**b**) at approximately three-fourths of cell length L to maximize overlap with the electric field of the microwaves with wavelength Λ. A circulator separates incident from the reflected microwaves. Taken from Ref. 10 with permission of the American Chemical Society.

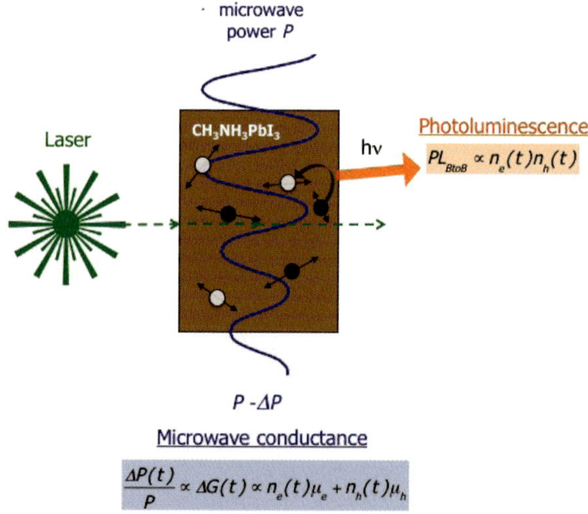

Scheme 1. Representation of TRPL and TRMC measurements on a thin film of MAPbI$_3$. In both techniques, electrons (closed circles) are excited to the CB by a short laser pulse, leaving mobile holes (open circles) in the VB. TRMC is used to measure the photoconductance (ΔG), which scales with the time-dependent concentration and mobility, μ of free electrons and holes. The blue sinusoidal line represents the magnitude of the microwave electric field as it passes through the sample. Radiative recombination of these mobile electrons and holes is probed by TRPL, which is a function of the concentrations of electrons ($n_e(t)$) and holes ($n_h(t)$).

Figure 3 shows a representation of the TRMC setup. Photo-excitation is realized by a laser pulses of 3–5 ns FWHM with a tunable wavelength at a repetition rate of 10 Hz. The maximum light intensity is on the order of 10^{15} photons/cm^2 per pulse. Metallic, neutral density filters with different optical densities are placed in between the laser and the sample to vary the photon fluence and thus the concentration of photoinduced charges. Monochromatic microwaves with a frequency in the range of 8.2–12.4 GHz are generated using a voltage-controlled oscillator. The sample of interest is placed in a microwave cell that ends with a metal grating (see Figure 3b), which fully reflects the microwaves. This cell is made from a gold-plated X-band waveguide. The sample is placed at three-fourths of the cell length Λ, so that its position corresponds to the maximum electric field strength for microwaves with wavelength Λ. A quartz window is glued on the top of the grating to seal the cell and avoid air exposure of the sample. The millimeter-sized openings of the grating transmit approximately 80% of the laser light. Most importantly, if the laser pulse induces free charges in the material, its conductivity increases and the total microwave reflection is reduced (ΔP). The cell is connected to a microwave source and detector *via* X-band waveguides (Figure 3a). A microwave circulator is incorporated to separate the incident from the reflected microwaves. The diode detector converts the microwave power into a direct current (DC), generating a voltage of 0.5–1.0 V when dropped across a resistor. Typically, ΔP is several orders of magnitude smaller than P. Therefore, an offset regulator is used to subtract the DC part, which leaves the laser induced AC signal undisturbed. This AC signal is then amplified (26 ×) and stored as function of time using a digital oscilloscope with a sampling rate of 4 GHz (every 0.25 ns). The trigger input of the oscilloscope is connected to a fast optical sensor that is illuminated by each laser pulse to start the acquisition of a microwave trace. Typically, the TRMC traces are averaged over 10^2 pulses.

As mentioned earlier, the ΔG is directly calculated from $\Delta P/P$ using the sensitivity factor K. The magnitude of K depends on the dimensions of the microwave cell, the dielectric properties of all the media in the cell and the microwave frequency. To determine K for a given sample, the following approach is used. The microwave reflection of a loaded microwave cell, P_M is calculated by numerically solving the Maxwell equations, using the characteristics of the cell and the dielectric properties of the media inside (sample, substrate, and gas). Then the microwave reflection P_M' of the sample in the same cell is calculated; however, the sample has now a small specific conductivity, σ such that $\Delta G < 1 \times 10^{-5}$ S. By definition, a change in conductance is related to the change in conductivity according to:

$$\Delta G = \frac{\Delta \sigma A}{d} = \frac{\Delta \sigma\, aL}{b} = \Delta \sigma\, \beta L \qquad (9)$$

Here, A is the area of the sample perpendicular to the electric field vector of the microwaves E_{mic} and d is the width of the sample in the direction of E_{mic}. For a sample with thickness L, A is given by $a \times L$, where a is the sample height. Similarly, d corresponds to the sample width b. Replacing the ratio of a and b by β, the change in conductance can now be calculated from the change in conductivity, the inner dimensions of the cell, β and the sample thickness, L.

The K factor can now be found by:

$$\frac{\Delta P_M}{P_M} = -k\Delta G$$

$$\frac{\left(P_M' - P_M\right)}{P_M} = -k\Delta \sigma \beta L \quad k = -\frac{\Delta \sigma \beta L}{\dfrac{\left(P_M' - P_M\right)}{P_M}} \qquad (10)$$

Important to note here is that K is determined for a specific combination of sample, microwave cell and frequency. To illustrate the latter dependency of the K factor, time-dependent $\Delta P(t)/P$ traces are measured at different frequencies within the range of 8.2–12.2 GHz using a cell filled with N_2 and a 1-mm thick quartz substrate covered with 250 nm MAPbI$_3$. Note that for each frequency, the used laser intensity is the same and hence the ΔG is constant. The $-\Delta P_{MAX}/P$ values are plotted versus frequency in Figure 4. A clear broad maximum can be discerned around 8.5 GHz. Hence, at this frequency K is maximum. This can be understood from the maximum overlap of the sample with the electric field. At

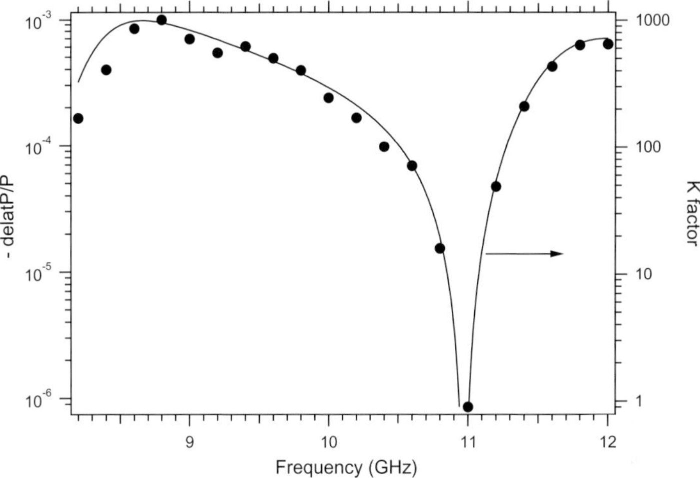

Figure 4. Dots: $-\Delta P_{max}/P$ as function of microwave frequency (8–12 GHz) for a 250-nm thin MAPbI$_3$ film. The solid line represents the frequency-dependent K factor, obtained from fitting the experimental data points.

11 GHz, the opposite is true: the electric field strength of the microwaves has a node here so the size of the overlap between field strength and sample is minimal. The full line gives the calculated frequency-dependent K factor for this sample.

Once K is known, ΔG can be obtained from the measured $\Delta P/P$ using eq 8. The TRMC signal can be expressed in the remaining two unknown parameters: that is, the mobility and the charge carrier yield, φ. Assuming that every absorbed photon creates a single positive and a negative charge carrier, eq 7 simplifies to:

$$\Delta\sigma = en(\mu_e + \mu_h) \tag{11}$$

in which μ_e is the electron mobility and μ_h is the hole mobility. The yield, φ, can be defined as

$$\varphi = \frac{Ln}{F_A I_o} \tag{12}$$

in which I_0 is the intensity of the laser in photons/pulse/unit area and F_A the fraction of light absorbed at the excitation wavelength. Combining eqs 9, 11, and 12, the product of yield and mobility is obtained from ΔG_{max}.

$$\varphi\left(\mu_e + \mu_h\right) = \frac{Ln}{F_A I_0}\frac{\Delta\sigma}{en} = \frac{L}{F_A I_0}\frac{\Delta G}{e\beta L} = \frac{\Delta G}{F_A I_0 e\beta} \tag{13}$$

Important to note here is that the actual thickness of the sample falls out, as long as the thickness, L is approximately less than a micrometer. This is especially important for samples for which a nonhomogeneous charge carrier profile is created within the sample upon optical excitation. Expressing the TRMC signal in the product of $\varphi(\mu_e + \mu_h)$ enables direct comparison between TRMC measurements from different samples.

The sensitivity of the TRMC setup can be further increased by partially closing the cell using an iris (Figure 3b). In this case, the cell acts as a resonant cavity for microwaves with wavelength Λ. The standing wave pattern in the cavity leads to more interaction with the sample, thereby enhancing K and enabling the use of lower photon fluences, however at the expense of a decreasing time resolution. For the *cavity*, the instrumental response time is 18 ns, while this is only 3 ns when a measurement is performed with the *open cell*. Figure 5 shows the effect of the instrumental response function (IRF) on the measured trace for a mathematically short (left) and long (right) photoconductance signal.

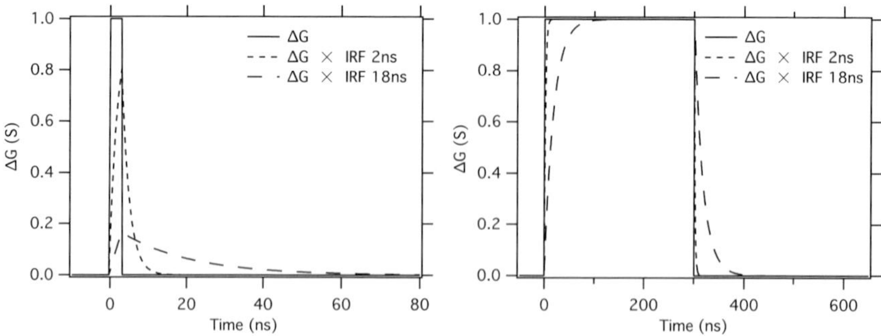

Figure 5. Comparison of the IRFs for the open cell (2 ns) and the cavity (18 ns) for a (left) short (3.5 ns) ΔG and (right) long (300 ns) ΔG. Taken from Ref. 42 with permission from the American Chemical Society.

D. Modeling of Kinetic Data

In order to extract quantitative data out of time-dependent PL and TRMC measurements, the following generic kinetic model can be used, as detailed in Scheme 2, to describe the charge carrier dynamics in OMHPs.[10,12,13] This model accounts for different recombination pathways of photoexcited electrons and holes as function of their concentrations. This model is based on a homogeneous generation of charges, which can be experimentally realized by using an excitation wavelength close to the absorption onset (see Figure 1). The initial number of photoexcitations n (cm^{-3}) depends on the light intensity of the laser I_0 (number of photons/cm^2) and the optical absorption at the excitation wavelength (F_A):

$$n = \frac{I_0 \times F_A}{L} \tag{14}$$

where L is the film thickness (cm). If the sample is much larger than the penetration depth of light, $1/\alpha$ can be used instead of x. This represents the distance at which 63% of the photons is absorbed, that is, where the light intensity is only $1/e$ of its original value.

Each absorbed photon initially forms an electron-hole pair. Depending on their binding energy, electron-hole pairs can dissociate into free CB electrons and VB holes. Note that only free, mobile charges contribute to the real photoconductance. Considering that the exciton binding energy of methylammonium lead iodide perovskites is only a few millielectronvolts,[14] the thermal energy at room temperature (i.e., ~26 meV) is sufficient to dissociate the majority of excitons into free charges. Therefore, it is assumed that at room temperature, every absorbed photon initially leads to one free CB electron and one free VB hole.

The time-dependent generation of charge carriers is denoted by $G_C(t)$, which takes into account the temporal profile and total light intensity of the laser pulse. The concentration of photoexcited CB electrons n_e and VB holes n_h are initially equal, but can be different as function of time depending on their recombination pathways. For instance, there could be intra-band gap states acting as electron traps. In this kinetic model, see

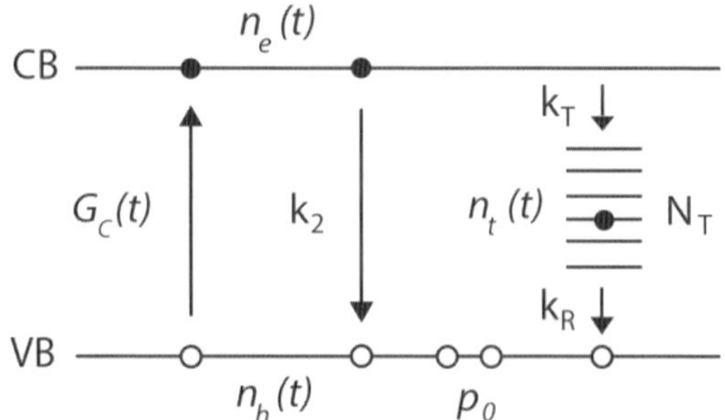

Scheme 2. Kinetic model of processes occurring upon photoexcitation of an OMHP, adapted from Stranks *et al.*[13] Here, $G_C(t)$ represents the photoexcitation of electrons (closed circles) from the VB to the CB. The electrons in the CB can recombine with holes (open circles) in the VB *via* k_2. In competition with k_2, electrons can be immobilized in intra-band gap trap states (total density N_T) *via* k_T. Finally, the trapped electrons can recombine with holes from the VB *via* k_R. In the case of a nonintrinsic OMHP, there will be additional holes (p_0, p-type) on the top of the photogenerated holes. Note that this fully mathematical model also holds for the opposite situation, i.e., an OMHP with trap states for holes and additional dark CB electrons (n_0, n-type). Taken from Ref. 15 with permission of American Chemical Society.

Scheme 2, the total trap density is represented by N_T. The concentration of trapped electrons is denoted as n_t. In a perfectly intrinsic semiconductor, the initial concentrations of CB electrons and VB holes upon photoexcitation are the same. However, in general, semiconductors are often unintentionally extrinsic due to impurities in the crystal lattice. In this case, there are additional CB electrons (n-type) or VB holes (p-type) already present before photoexcitation. In Scheme 2, the concentration of thermal equilibrium charges (dark carriers) is represented by p_0. Note that p_0 does not contribute to the photoconductance. However, the recombination of photogenerated charges is affected by p_0, because the total concentration of VB holes ($n_h + p_0$) is larger than the concentration of CB electrons (n_e).

The following set of differential eqs (15–17) implements Scheme 2 and describes n_e (16) n_h (17), and n_t (18) as function of time. The rate constants for band-to-band electron-hole recombination, trap filling and trap emptying are represented by k_2, k_T, and k_R, respectively (Scheme 2). These differential equations are coupled: each photoexcited electron eventually decays back to the ground state.

$$\frac{dn_e}{dt} = G_C - k_2 n_e (n_h + p_0) - k_T n_e (N_T - n_t) \tag{15}$$

$$\frac{dn_h}{dt} = G_C - k_2 n_e (n_h + p_0) - k_R n_t (n_h + p_0) \tag{16}$$

$$\frac{dn_t}{dt} = k_T n_e (N_T - n_t) - k_R n_t (n_h + p_0) \tag{17}$$

As detailed in eq 15, the decrease of n_e over time depends on the recombination rate with VB holes $k_2 n_e (n_h + p_0)$ and trapping rate $k_T n_e$ $(N_T - n_t)$, where $N_T - n_t$ is the concentration of available traps. Simultaneously, n_t increases due to trapping and decreases depending on the recombination rate with holes $k_R n_t (n_h + p_0)$. Solving the equations using numerical methods yields the time-dependent concentrations of n_e, n_h, and n_t.

The change in photoconductance as function of time is calculated from the product of charge carrier concentrations and mobilities according to:

$$\Delta G(t) = e\left(n_e(t)\mu_e + n_h(t)\mu_h\right)\beta L \tag{18}$$

In which L is the thickness of the film, μ_e and μ_h the electron and hole mobilities, respectively, and β is the ratio of the inner dimensions of the microwave cell. Here, the trapped charge carriers, n_t are immobile and do not contribute to ΔG. The mobilities are assumed to be constant within the time window of the measurement and independent of the charge density. Finally, a convolution is applied to take into account the IRF of the setup and model the experimental TRMC traces.[10]

Assuming that PL originates only from second order band-to-band recombination of electrons and holes, the PL lifetime is proportional to the concentration of CB electrons, VB holes, and k_2. The TRPL can then be calculated according to:

$$PL(t) = \frac{k_2 n_e(t)(n_h(t) + p_0)L}{I_0 F_A} \tag{19}$$

where the same set of differential equations is solved. Only the generation term G_C is altered due to the fact that a different laser with a shorter pulse and wavelength was used. This kinetic model can also be used to analyze time-resolved data obtained from different techniques, for example, THz or transient absorption (TA) spectroscopy, or describe charge carrier dynamics in other semiconductor materials.

III. Transient Photoconductivity Measurements in the Picosecond to Few Nanosecond Time Scales

A. Properties of Intrinsic MAPbI$_3$ Material

Figure 6 shows the transient photoconductivity kinetics of neat MAPbI$_3$ film and MAPbI$_3$ attached to mesoporous nanoparticles of aluminum oxide (MAPbI$_3$/Al$_2$O$_3$) and titanium oxide (MAPbI$_3$/TiO$_2$). The first 4 ps of the transient photoconductivity, normalized to 1, is shown in Figure 6a. It can be observed that the rise time of MAPbI$_3$ and MAPbI$_3$/Al$_2$O$_3$ are similar, that is, characterized by an ultrafast instrument-limited rise, about 70% of the total signal amplitude, followed by 2–3 ps rise, about 30% of the total amplitude. As described in the methods section, $\Delta\sigma$ is a product of charge concentration and mobility. This allows two possible interpretations of the rise time behavior. First, that the ultrafast component is generation of charged species while the ps component is the dissociation time of these species resulting to higher mobility. The exciton binding energy MAPbI$_3$ is in the order of tens of millielectronvolts.[16–18] This means that upon photoexcitation, the charged species should be loosely bound excitons. Considering that at room temperature, $k_B T$ (~26 meV) is on the same order of magnitude as the exciton binding energy, charges dissociate easily in 2–3 ps gaining mobility. This is then manifested in the

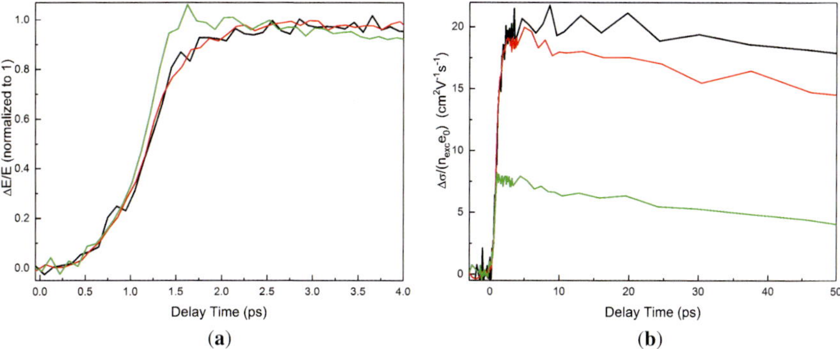

Figure 6. Transient photoconductivity of neat $MAPbI_3$ (black), $MAPbI_3/Al_2O_3$ (red), and $MAPbI_3/TiO_2$ (green) (**a**) normalized to 1 up to 4 ps and (**b**) normalized to excitation density $n_{exc}e$ up to 50 ps. λ_{pump} = 400 nm, I_{exc} = 1.7×10^{13} ph/cm^2 per pulse. Taken from Ref. 21 with permission of the American Chemical Society.

photoconductivity kinetics as an additional rise. The second possibility is that upon photoexcitation both highly mobile charges (70%) and loosely bound excitons (30%) are simultaneously generated. This requires however a heterogeneity in the exciton binding energy. According to Lin et al.,[17] the exciton binding energy is as little as 4 meV while for Sheng et al.[18] it can be as high as 50 meV. It is likely that this particular sample has different exciton binding energies as well depending on the quality of the film at different areas. Note that this two-step rise behavior is common for the two samples showing that the presence of Al_2O_3 nanoparticles does not alter the early time dynamics of the charges.

The transient photoconductivity of perovskite attached to TiO_2 ($MAPbI_3/TiO_2$) is quite different from the two previous samples as it only has one ultrafast instrument-limited rise. As both electrons and holes are generated in the perovskite, the favorable band alignment between the CBs of perovskite and TiO_2 allows electrons to be transferred. Similar rise times have been observed in Ru-N3 dye attached to TiO_2 nanoparticles[19] as well as in CdSe quantum dot attached to TiO_2 nanoparticles.[20] Therefore, this instrument-limited rise in the transient photoconductivity kinetics was assigned to ultrafast electron injection from perovskite into TiO_2. A more detailed discussion on electron and hole injection from perovskite material to metal oxide will be presented later in this section.

To further understand the behavior of charge carriers in the picosecond time scale, the transient photoconductivity was normalized to excitation density using eq 1. The plot in Figure 6b shows the resulting mobility. Again, it should be noted that the mobility calculated here is based on the assumption that at the earliest time scale all photons absorbed are converted to charged species, that is, quantum yield φ is 1. Therefore, the mobility shown here is a lower estimate. Based on this assumption, the mobility obtained from neat $MAPbI_3$ and $MAPbI_3/Al_2O_3$ is 20 cm^2/Vs, while for $MAPbI_3/TiO_2$ the mobility is 7.5 cm^2/Vs. Hendry *et al.* reported that the mobility of electrons in TiO_2 is 0.1 cm^2/Vs,[22] which is negligibly low compared to the mobility in neat $MAPbI_3$. As discussed earlier, there is an ultrafast injection of electrons from $MAPbI_3$ to TiO_2. This means that 7.5 cm^2/Vs mobility calculated in $MAPbI_3/TiO_2$ should be coming from holes left in the perovskite because the electrons that are transferred to TiO_2 barely contribute to the signal. As a result, the 20 cm^2/Vs mobility in the neat $MAPbI_3$ and $MAPbI_3/Al_2O_3$, where both electrons and holes stay in the perovskite means that 12.5 cm^2/Vs should be coming from electrons because the mobility of holes is 7.5 cm^2/Vs. These have several direct implications in the operation of $MAPbI_3$-based solar cells. (1) The mobility of electrons and holes are at least 100 times higher than in organic solar cells (μ_e = 0.005 cm^2/Vs, μ_h = 0.02 cm^2/Vs)[23] allowing a faster and more efficient extraction of charges to the electrodes. (2) The difference in the mobility between electrons and holes is within a factor of two. Because of this, electrons reach the electrodes as soon as the holes reach the counter electrode, that is, the transport is balanced. The latter prohibits building up of space charge–limited current that lowers the PCE.[24] This result also agrees with the estimation of the effective masses of electrons and holes.[25] The value of mobility and the apparent decay of the transient photoconductivity of neat $MAPbI_3$ and $MAPbI_3/Al_2O_3$ are very similar (Figure 6b). This reinforces the assertion that Al_2O_3 nanoparticles do not influence the charge dynamics. Rather, it acts like an inert scaffold resulting in a more uniform morphology of the film.[21]

To study the mode of transport of charges in $MAPbI_3$, the transient photoconductivity spectra were obtained for the three different samples. In the plot of Figure 7a, the real part of the photoconductivity is positive while the imaginary part is negative at 10 ps. This is true for all three samples except for the lower value of the real part of photoconductivity of MAPbI$_3$/TiO$_2$, which is dominated by the holes. For MAPbI$_3$/Al$_2$O$_3$, the photoconductivity spectra were also obtained at different pump-probe delays up to 1 ns. The shape of the photoconductivity spectra is reminiscent of our previous results in organic,[23] dye-sensitized,[19] and quantum-dot sensitized[20] solar cells, where the mode of transport is restricted or confined, that is, positive real part and negative imaginary part of photoconductivity similar to that shown in Figure 7a. In other words, despite its high mobility, once a charge hits a barrier or boundary it will not continue on its path but

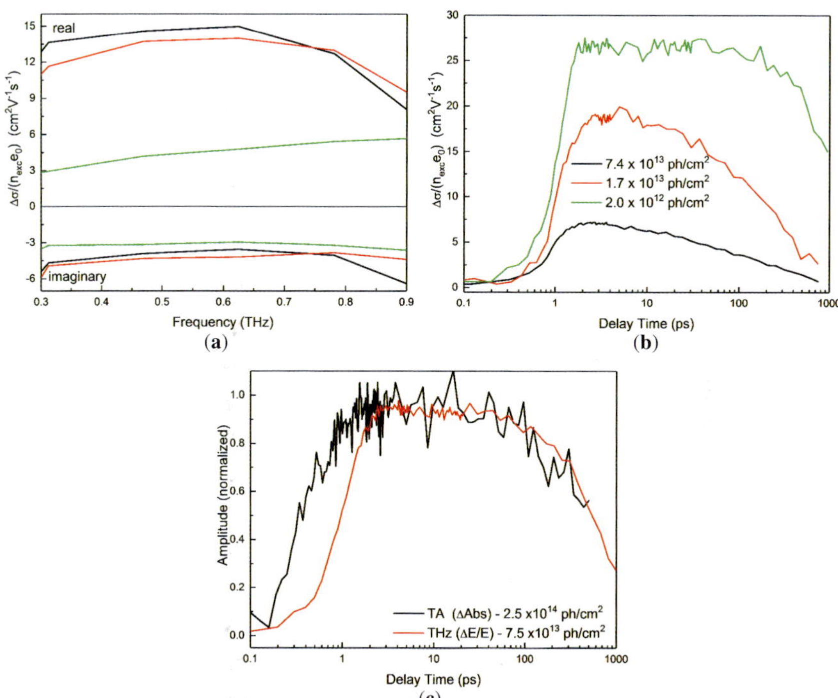

Figure 7. (a) THz photoconductivity spectra measured 10 ps (trace with symbols) after photoexcitation and normalized with $n_{exc}e$. Solid traces are THz photoconductivity spectra of MAPbI$_3$/Al$_2$O$_3$ at different pump probe delays (blue 100 ps, cyan 200 ps, magenta 600 ps, gray 950 ps). (b) Intensity dependence THz kinetics of MAPbI$_3$/Al$_2$O$_3$. Recombination commences after 200 ps for lowest excitation density where the highest mobility is obtained. (c) Comparison of TA and THz kinetics for neat MAPbI$_3$ showing that THz mobility remains constant for at least 1 ns. Taken from Ref. 21 with permission of the American Chemical Society.

rather backscatter or recombine or get trapped to that boundary/barrier. The dependence of the transient photoconductivity to the excitation intensity is shown in Figure 7b. At highest excitation fluence (7.4×10^{13} ph/cm^2), the mobility is 5 cm^2/Vs and decays rapidly from 3 ps. On the other hand, at about 40 times lower excitation fluence (2.0×10^{12} ph/cm^2), the mobility is 25 cm^2/Vs and the onset of decay is extended up to 200 ps. This means that the excitation intensity induces an additional mechanism that forces the mobility to be low and accelerates the decay. This is similar to previous results obtained for organic solar cells, where the transient photoconductivity decays in a few picosecond at high pump intensity, but extends its lifetime to at least half a nanosecond when the fluence is lowered by a factor of 100.[26] In the same manner, it is surmised that the lower mobility and fast decay here is due to charge pair annihilation, that is, at the highest excitation intensity, charge carriers are very close to each other and will result in recombination. At low intensities, charges are sparse and the possibility of recombination is less. To verify this, the photoconductivity kinetics were compared to those observed with transient absorption measurements (Figure 7c). After a few picoseconds, the two traces overlap until 1 ns. Transient photoconductivity is the product of charge population and mobility while transient absorption directly monitors solely the change in charge population. Because the two traces overlap, this means that the decay in transient photoconductivity should be emanating from the change in charge population only, at a rate shown by the transient absorption decay. Consequently, this means that the mobility remains constant up to at least 1 ns otherwise the decay rate should be faster.

In summary, the characteristics of MAPbI$_3$ are as follows: the primary photophysical products are highly mobile charges and loosely bound excitons whose formation ratio depends on the quality of the film; electrons are injected into TiO$_2$ on a sub-picosecond time scale; and charge mobilities are balanced, are at least two orders of magnitude higher than in organic solar cell materials and are maintained for at least 1 ns.

B. Properties of Intrinsic FAPbI$_3$ Material

Despite the almost ideal solar cell properties of MAPbI$_3$, the methyl ammonium cation (MA$^+$) that is in the center of the unit cell spins erratically. The collective realignment of MA^+ tends to screen a device's built-in potential that could reduce its photovoltaic performance.[27] One possible improvement is to find another cation with a smaller dipole moment. It has been reported that this is the case in formamidinium lead triiodide (FAPbI$_3$) whose PCE has demonstrated to reach over 20%.[28] The superior performance of this material is shown to be due to the broader solar spectrum and higher mobility of the charge carriers.[29,30] The broader absorption spectrum that lowers the band gap, that is, from 1.54 to 1.47 eV, is due to the increase in the effective cation radius from 217 pm in methylammonium to 253 pm in formamidinium.[31] These changes in its physical properties will now be compared to its photophysical properties as probed by time-resolved THz spectroscopy.

Plotted in Figure 8a are the transient photoconductivity spectra of FAPbI$_3$ film excited at 600 and 760 nm, taken 20 ps after photoexcitation and normalized to excitation density. The resulting mobility is 60–80 cm^2/Vs; about three to four times higher than that of MAPbI$_3$. The higher apparent mobility can be explained as due to the expansion of the lead iodide lattice that could result in longer diffusion lengths.[29] In fact, the diffusion length

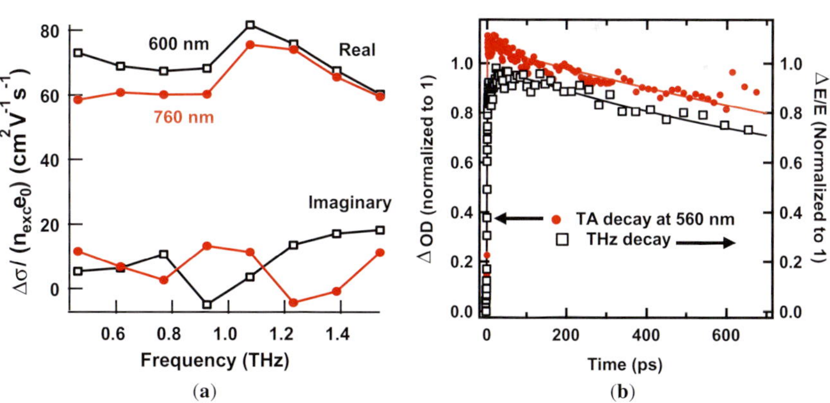

Figure 8. (a) Transient photoconductivity spectra of FAPbI$_3$ film excited at 600 (open squares) and 760 nm (solid circles) using a fluence of 5×10^{12} photons/cm^2. (b) Transient absorption and transient photoconductivity kinetics normalized to 1. The solid lines are from multi-exponential fits to guide the eye. For clarity, the TA decay signal was offsetted. Taken from Ref. 32 with permission of the American Chemical Society.

reported for MAPbI$_3$ is about 3.4 μm[30] only while calculations based on the data shown in Figure 8 suggest that for FAPbI$_3$ the diffusion length is about 30 to 25 μm.[32] Moreover, it was observed that better quality and more continuous film morphology is easier to obtain here than in MAPbI$_3$. Note that the real part and imaginary part of the photoconductivity spectra are both positive. This is an indication of a long-range mode of transport (Drude-like), that is, charges are moving in an electron gas-like manner.[33,34] This reaffirms the result that the diffusion lengths are longer and the quality of the film is better. In contrast, MAPbI$_3$ has shorter diffusion lengths forcing the charge carriers to be restricted or confined as discussed earlier. In Figure 8b, the photoconductivity of FAPbI$_3$ is compared to the transient absorption. It is clear that both traces have the same decay rate, which resembles Figure 7c of MAPbI$_3$. Ergo, the mobility of carriers in FAPbI$_3$ is constant for at least 1 ns.

The first 100 ps of the transient photoconductivity kinetics of FAPbI$_3$ pumped at four different energies are shown in Figure 9a. For wavelengths longer than 500 nm, there is a two-step rise similar to MAPbI$_3$. However, in this case, the second step rise is more subtle, that is, only about 5–10% of the total signal amplitude in contrast with MAPbI$_3$ where the second step is about 30% (Figure 6a). In the same manner, the origin of this two-step rise can be explained as due to the heterogeneity of exciton binding energy. Because the second step rise is just 10%, it can be surmised that the exciton binding energy is lower than 30 meV, probably closer to 4 meV.[17] Interestingly, at 500 nm, the second step rise is absent. This means that at this energy, that is, 400 meV higher energy than 600 nm, all photogenerated carriers are directly converted to highly mobile charges due to this excess energy.

The transient photoconductivity of FAPbI$_3$ is further monitored up to 8 ns. Plotted in Figure 9b is the resulting mobility at three excitation fluences. At the earliest time scale, the initial mobility is 60 cm^2/Vs for all three excitation conditions. This is in contrast to the results of MAPbI$_3$ (Figure 7b), where the mobility is lower at higher excitation and higher mobility at lower excitation. This means that the charge pair annihilation is not present at these conditions. In other words, although the high excitation density results into very high concentration of charge carriers, the photogenerated charges can still diffuse and pull away from each other.

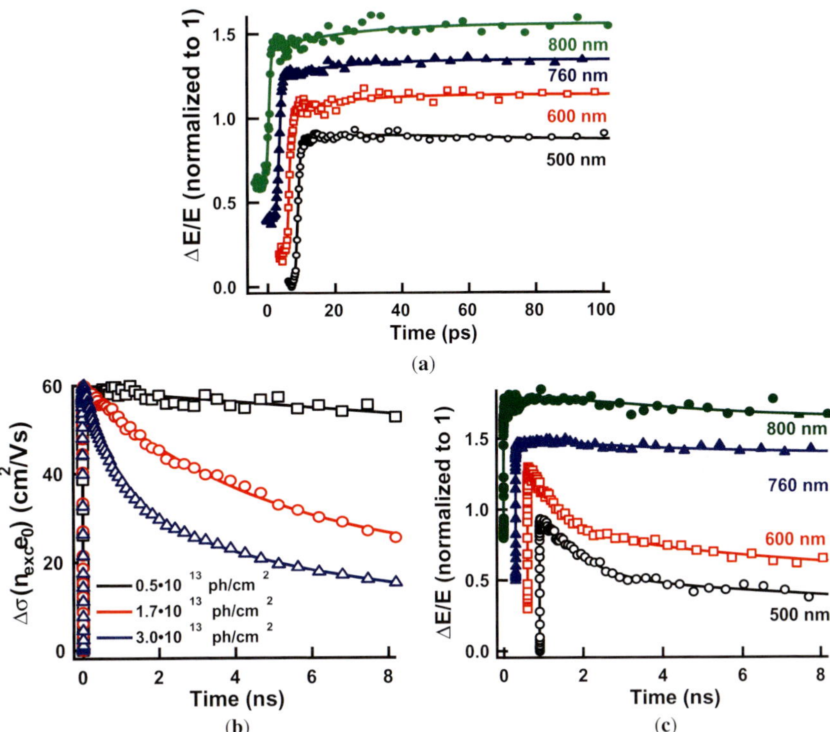

Figure 9. Transient photoconductivity of FAPbI$_3$ excited (**a**) using 500, 600, 760, and 800 nm at a pump fluence of 5×10^{12} ph/cm^2 at first 100 ps; (**b**) using 760 nm at varying fluences of 5×10^{12} ph/cm^2 (squares), 1.7×10^{13} ph/cm^2 (circles), and 3.0×10^{13} ph/cm^2 (triangles); and (**c**) similar to (a) but up to 8 ns. Taken from Ref. 32 with permission of the American Chemical Society.

This is consistent with the report of Eperon *et al.* that substituting FA$^+$ to MA$^+$ will expand the lattice of FAPbI$_3$.[29] This expansion in lattice allows wider distribution of charge carriers. At the later time scale and at lowest excitation intensity, there is a minute to almost no decay in the transient photoconductivity. This shows that neither the concentration of the charges nor the mobility has decayed in 8 ns. This again reinforces the favorable expanded lattice and higher quality of film in FAPbI$_3$. Second order geminate recombination is seen as the cause of the intensity-dependent decay at higher excitation in Figure 6b. At relatively low fluence of 5×10^{12} ph/cm^2, the role of the excitation energy to the decay dynamics of charge carries was investigated. This is plotted in Figure 9c where four excitation energies 500, 600, 760, and 800 nm were used up to 8 ns. At 760 and 800 nm, the photoconductivity kinetics are nearly flat. However, for 500 and 600 nm the decay starts to be significant. One should note that the absorption coefficient of the sample at these two wavelengths is about three times higher compared to longer wavelength (760 and 800 nm).[32] As a result, the density of photogenerated charges, although maintained at the same intensity, would be three times higher. The outcome would be similar to the second order nongeminate recombination where the probability of charges to meet are higher at elevated population.

In summary, thin film FAPbI$_3$ is a promising perovskite solar cell material. Its expanded lattice results in higher mobilities, longer diffusion lengths, Drude-like behavior of charge carriers, and sustained mobility up to at least 8 ns.

C. Charge Injection to Metal Oxide Electrodes

When MAPbI$_3$ was first employed in a Graetzel-type solar cell device in 2009, liquid electrolyte of iodide was used and the resulting PCE was about 3.8%.[1] However, the device was not stable because the perovskite instantly dissolved in the highly polar liquid electrolyte. Since then, there have been efforts to find substitutes and one of the promising candidates is nickel oxide (NiO). This is due to its high transparency in the visible region as well as its high chemical stability. In 2015, a solvent-engineered perovskite solar cell device that used a NiO as counter electrode reached a PCE of 17.3%.[35] Moreover, by using p-type NiO and n-type ZnO (or TiO$_2$) as electrodes in an all-metal oxide charge extraction layer device, the group of Che *et al.* and You *et al.* demonstrated that efficiencies of at least 15% can be reached with long-term stability.[36,37] In this context, it is interesting to understand the mechanism and time scale of injection of electrons and holes in metal oxide electrodes.

Plotted in Figure 10a are the transient photoconductivity kinetics of neat MAPbI$_3$ film and MAPbI$_3$ attached to metal oxide nanoparticles of zirconium oxide (ZrO$_2$), TiO$_2$, and NiO for the first 100 ps. For neat film

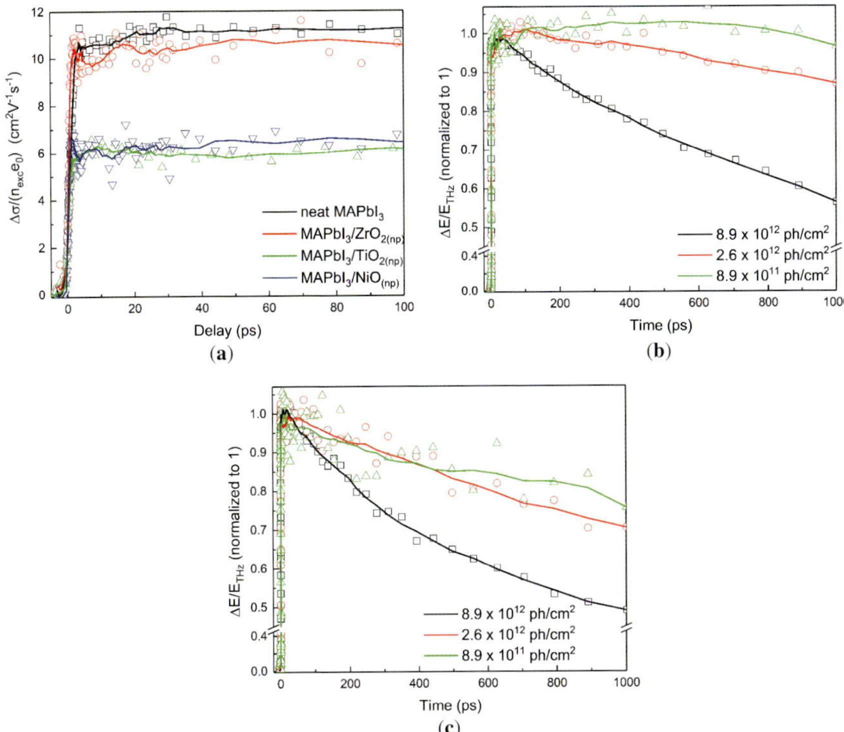

Figure 10. THz photoconductivity kinetics of (**a**) neat $MAPbI_3$ (black), attached to nanoparticles of ZrO_2 (red), TiO_2 (green), or NiO (blue) normalized to excitation density $n_{exc}e$ (λ_{pump} = 400 nm, $I_{exc} = 1.7 \times 10^{12}$ ph/cm^2 per pulse) for the first 100 ps, while (**b**) neat $MAPbI_3$ film and (**c**) $MAPbI_3$/NiO are pumped at three different intensities normalized to 1 and up to 1 ns. Open symbol are data while solid trace is guide for the eye. Taken from Ref. 41 with permission of the American Chemical Society.

of $MAPbI_3$ where both electrons and holes are generated and stay in the perovskite material, the mobility obtained is 12 cm^2/Vs. This is just about half of the mobility measured in the previous perovskite sample (20 cm^2/Vs, Figure 3b). This can be possibly explained by the minute differences in the preparation routes. It was reported that despite the similarity in the preparation protocol, parameters such as thermal annealing time have a major influence to the type and concentration of defects that a perovskite material will have.[38] As a result of this, the quantum yield may change which cannot be estimated by this technique alone. Moreover, depending on the morphology of the resulting film, saturation in the absorption spectra may be observed[39] that could complicate the accurate normalization of the photoconductivity data. More accurate comparison of mobility may be obtained when samples from the same batch are measured because the preparation conditions are identical. In any case, all of the past measurements routinely yielded a mobility from 10 to 25 cm^2/Vs whose deviation is rather small to affect the interpretation of the photophysical properties in the results. Right after the instrument-limited rise in the photoconductivity of neat $MAPbI_3$, there seems to be a 10–20% additional rise that lasts up to about 40 ps before plateauing. This could indicate loosely bound excitons with binding energies slightly higher than those mobile charges that are generated instantaneously. However, within the signal-to-noise of the instrument and at this very low excitation condition, this cannot be unambiguously assigned. When $MAPbI_3$ is attached to ZrO_2, the mobility is similar to the neat $MAPbI_3$. The CB of ZrO_2 nanoparticles ($E_g = 5.34$ eV) is quite high with respect to the CB of $MAPbI_3$ which therefore prohibits electron injection. This means that both electrons and holes remain in the perovskite material with mobilities similar to the neat $MAPbI_3$ film, justifying the same mobility of 12 cm^2/Vs. Thus, the ZrO_2 nanoparticle serves as a scaffold, similar to the role Al_2O_3 plays.[21] For $MAPbI_3$/TiO_2, the mobility measured is 6 cm^2/Vs, half of the mobility in the neat $MAPbI_3$. As discussed earlier (Figure 7), due to the favorable CB alignment at the interface, the reduction in total mobility is due to ultrafast electron injection. In order words, electrons are immediately removed from perovskite and transferred into TiO_2 and the mobility obtained is due to holes that stay in the perovskite. Interestingly, the $MAPbI_3$/NiO sample has very similar transient photoconductivity kinetics and also the mobility is 6 cm^2/Vs. This means that one of the charge carriers (electrons or holes) disappears at the earliest time scale. There is a 0.2 eV offset in the VB energies between the perovskite and NiO. Moreover, the intrinsic conductivity of stoichiometric NiO is 10^{-13} S/cm.[40] All of these strongly suggest that there is an ultrafast hole injection from $MAPbI_3$ to NiO. In fact, a corresponding solar cell device was fabricated in the same batch as the films used for THz measurement. A PCE of 13.7% was obtained confirming that holes should have been transferred from perovskite to NiO counter electrode and eventually extracted to provide current.[41] Due to the very low mobility of charges in NiO, injected holes are not detected by the instrument. Thus, the 6 cm^2/Vs mobility is solely originating from the electrons left in the perovskite material. This is very similar to the mobility of the holes left in the perovskite for the MAPbI$_3$/TiO$_2$. This affirms the previous conclusion (Figure 6b), that the mobilities of carriers in perovskite do not differ by more than a factor two, in fact, in this case, they are balanced.

To determine the interfacial recombination time of the injected charge carriers, the neat MAPbI$_3$ was pumped at three excitation fluences and measured up to 1 ns. The transient photoconductivity kinetics are shown in Figure 10b, normalized to 1 to highlight the decay rate. At the highest excitation intensity (8.9 $\times$ 10^{12} ph/cm^2), the photoconductivity has reduced to 50% in 1 ns. Using a fluence of 2.5 $\times$ 10^{12} ph/cm^2, the decay slowed and is just 10% less than its initial value. For the lowest excitation (8.9 $\times$ 10^{11} ph/cm^2), the kinetics are flat up to 1 ns. There seems to be an apparent rise in hundreds of picoseconds. However, it should be noted that the fluence used here is low, that is, about 100- to 1000-fold lower than those used previously used in organic solar cells.[26] As such, similar to the apparent rise in Figure 7a, this could be within the signal to noise of the instrument and its assignment cannot be straightforwardly ascertained. However, one can surmise that at the lowest excitation condition, neither the population nor the mobility changes, while at higher pump fluences the decay is due to second order recombination.[41] This has important consequences in understanding the mechanism of recombination in MAPbI$_3$/NiO. As shown in Figure 10c, MAPbI$_3$/NiO is also pumped at three excitation intensities similar to neat MAPbI$_3$. At the highest excitation, the decay is very similar to the decay of neat MAPbI$_3$. However, for fluence of 2.5 $\times$ 10^{12} ph/cm^2, the decay is faster, having 80% of its initial amplitude while for neat MAPbI$_3$ (at similar fluence) the photoconductivity is still 90% of its initial value. At the lowest excitation, the photoconductivity of neat MAPbI$_3$ is flat while MAPbI$_3$/NiO have reduced to about 80%. Thus, the decay in MAPbI$_3$/NiO should not be due to the excitation intensity used but from some other decay channel. As presented earlier, there is an ultrafast hole injection. Once in NiO, the mobility is very low and therefore it takes time for it to diffuse away from the interface. On the other hand, electrons left in the perovskite are highly mobile. Thus, it is surmised that the rate at which the electrons find a hole that is immobilized at the interface, determines the recombination, in this case, manifested as decay in photoconductivity in hundreds of picoseconds to few nanoseconds time scale. One should not discount the possibility that the NiO nanoparticles are not defect-free. These defects act as electron traps, which can also be in the same time scale. The ratio, which of these two mechanisms is more dominant cannot be ascertained by this technique alone and further investigation on this is requested.

In summary, ultrafast electron and hole injection are confirmed from perovskite to metal oxide, TiO_2 and NiO, respectively. This is mainly due to the favorable band alignment with the energy bands of perovskite. However, the low mobilities of charges in these metal oxides hinders extraction to electrodes. Ways to increase their mobility by lowering surface defects may be one of the techniques to address this issue.

D. Charge Injection to Organic Electrodes

One of the main advantages of perovskite solar cells, similar to organic solar cells, is its easy manufacturability. Using standard kitchen chemistry protocols, one can prepare a solar cell device without high temperature curing and/or annealing. For this reason, the use of organic molecules as electrodes has been an attractive proposition where all the component layers are deposited on top of each other just by spin coating. This has led to extensive research efforts in using phenyl-C61-butyric acid methyl (PCBM) and Spiro-OMeTAD as electron and hole acceptor materials, respectively. Similar to the above results, THz radiation was used to determine the time scale and mechanism of charge transfer when these organic molecules are used as electrodes.

Plotted in Figure 11a are the transient photoconductivity kinetics of neat $MAPbI_3$, $MAPbI_3$/PCBM, and $MAPbI_3$/Spiro-OMeTAD for 1 ns and normalized to the excitation density. For this batch of samples, neat $MAPbI_3$ has a mobility of 15 cm^2/Vs. The difference in the mobility from two previous neat $MAPbI_3$ samples (Figures 7b and 10a) are discussed earlier and details can be found in Refs. 37 and 38. The photoconductivity

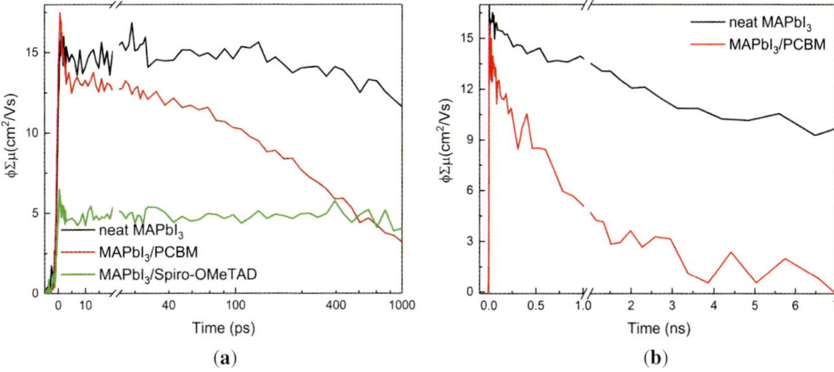

Figure 11.   Transient photoconductivity kinetics of (**a**) neat $MAPbI_3$, $MAPbI_3$/PCBM, and $MAPbI_3$/Spiro-OMeTAD for 1 ns and (**b**) $MAPbI_3$ and $MAPbI_3$/PCBM for 7 ns time window normalized to the excitation intensity of 2.1×10^{12} ph/cm^2 per pulse and 8.0×10^{12} ph/cm^2 per pulse, respectively (I_{pump} = 590 nm). Taken from Ref. 42 with permission of the American Chemical Society.

kinetics of neat $MAPbI_3$ remain flat for at least 1 ns showing that both the population of charge carriers and its mobility are constant at this time scale. Similar conclusions can be drawn from the trace of $MAPbI_3$/Spiro-OMeTAD, but its mobility is three times less (5 cm^2/Vs). Considering the energy alignment of the VBs at the interface of $MAPbI_3$ and Spiro-OMeTAD, there is a wide difference of 0.57 eV.[43] It is also by using Spiro-OMeTAD as hole transporting material that high PCEs are recorded for perovskite solar cells. These arguments strongly suggest that hole injection is efficient. The fact that the measured intrinsic conductivity of Spiro-OMeTAD is 10^{-8} S/cm[44] confirms that the reduction in mobility is due to disappearance of holes as these are injected to low-mobility Spiro-OMeTAD. Furthermore, because this reduction happens in the earliest time scale (instrument-limited), indicates that the hole injection is ultrafast. The mobility of 5 cm^2/Vs is the mobility of electrons left in the perovskite.

Interestingly, the transient photoconductivity kinetics of $MAPbI_3$/ PCBM exhibit a different behavior. At the earliest time scale, it has a mobility similar to neat $MAPbI_3$ but decays in hundreds of picoseconds reducing to a third of its original amplitude in 1 ns. Because it is measured at the same excitation fluence as neat $MAPbI_3$, where no decay is observed, it can be established that the decay is not due to second order recombination. Thus, there is another decay channel that induces such sub-nanosecond decay. At a longer time window (Figure 11b), the decay of MAPbI$_3$/PCBM is faster, that is, returning to zero value after 7 ns. This measurement is done in somewhat higher excitation intensity where second order recombination is present, as shown by the decay of neat MAPbI$_3$, but still the decay is faster than the neat MAPbI$_3$. The offset in the CBs between the perovskite and PCBM was reported to be 0.3 eV[45] which is just half of the difference in the VBs of perovskite and Spiro-OMeTAD. Due to this, it is possible that the time scale of electron injection is extended falling within sub-nanosecond to nanosecond time scale. Furthermore, similar to the arguments used in NiO hole injection, the low-mobility PCBM immobilizes injected electrons at the interface. Therefore, the rate at which the holes recombine with electrons at the interface can also explain the decay. It is surmised that the hundreds of picoseconds to few nanoseconds time scale photoconductivity decay is a convolution of slow injection convoluted with recombination between injected electrons in PCBM and holes left in perovskite.

Although it is favorable to use all organic extraction layers due to its ease in preparation, similar to metal oxides, the low mobility of these molecules remains to be a top concern. As charge carriers are pinned at its interface, recombination is hastened and the amount useful current extracted is lowered. Finding other high-conductivity materials or doping could retard interfacial recombination.

IV. Transient Photoconductivity Measurements in the Hundreds of Nanoseconds to Microsecond Time Scales

A. Thin Film Perovskites

The kinetic model presented in the methods section was used to describe TRMC and TRPL measurements on solution-processed thin films of MAPbI$_3$ and MAPbI$_{3-x}$Cl$_x$. Planar thin films were compared to their mesostructured analogues, that is, OMHPs infiltrated into an insulating Al$_2$O$_3$ scaffold, which both have been used to produce state-of-the-art perovskite solar cells.[46,47] The charge carrier mobilities and lifetimes were deduced from the measurements,[10] whereas global kinetic parameters were obtained from modeling both the TRMC and TRPL data for each sample.

Planar thin films (~300 nm) of $MAPbI_3$ and $MAPbI_{3-x}Cl_x$ were spin coated directly on quartz substrates using PbI_2 and $PbCl_2$ as the lead source, respectively.[48] The fraction of chloride (x) varies depending on processing conditions, but typically appears not to be more than a few hundred parts per million.[49–52] More importantly, scanning electron microscope (SEM) images showed that in the $MAPbI_{3-x}Cl_x$ thin film, the crystals are micrometer-sized, whereas those in $MAPbI_3$ are on the order of only a few hundred nanometers.[15] In addition, mesoporous alumina (Al_2O_3) scaffolds were infiltrated with OMHPs using the same precursor solutions, yielding mesostructured films (~400 nm thick) of $MAPbI_3$/ Al_2O_3 and $MAPbI_{3-x}Cl_x/Al_2O_3$. The concentrations of precursors were such that there was negligible formation of perovskite capping layer on the top of the Al_2O_3 scaffold, that is, the perovskite is fully contained within the mesoporous layer. The X-ray diffraction (XRD) patterns confirmed that all OMHP films are highly crystalline, showing strong reflections characteristic for the {110} planes of $MAPbI_3$-like perovskites. Using the Scherrer equation,[53] the crystal domain sizes are estimated to be approximately 50 nm for the mesostructured OMHPs.[15] Here, the Al_2O_3 scaffold limits the growth of OMHP crystals and hence the crystal domains are of similar size in $MAPbI_3/Al_2O_3$ and $MAPbI_{3-x}Cl_x/Al_2O_3$.

Figure 12a and b displays the fraction of absorbed light (F_A, solid lines) and corresponding PL (dashed lines) spectra of the different samples. For all samples, the maximum emission wavelengths are located at 770 ± 5 nm (1.6 eV).

The TRMC technique was used to measure ΔG as function of time after photoexcitation of the different samples at $\lambda = 600$ nm using incident light intensities ranging over four orders of magnitude: from 10^9 to 10^{13} photons/cm^2 (3×10^{-4} to 3 μJ/cm^2/pulse). Additionally, TRPL was performed to specifically gain insight in radiative band-to-band recombination (k_2 in Scheme 2). The kinetic model was then used to globally fit the TRMC and TRPL results, using a single set of kinetic parameters for each sample. The traces obtained from fitting the experimental data with the

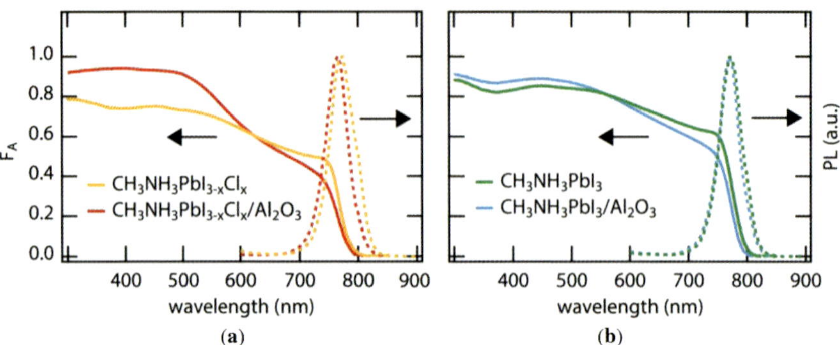

Figure 12. (**a** and **b**) Fraction of light absorbed (F_A) by OMHP films with structures as indicated (solid lines) and corresponding PL spectra normalized to 1 (dashed lines). Taken from Ref. 15 with permission of the American Chemical Society.

kinetic model from Scheme 2 are shown in the right panels of Figure 13. The kinetic parameters that were used to describe the measured TRMC and TRPL data for this specific sample are listed in Table 1.

The left panels in Figure 13 show ΔG normalized for absorbed number of photons after photoexcitation of $MAPbI_{3-x}Cl_x$ (a), $MAPbI_3$ (b), $MAPbI_{3-x}Cl_x/Al_2O_3$ (c) and $MAPbI_3/Al_2O_3$ (d). The rise of the signal originates from the formation of mobile charges; the maximum signal height represents the product of charge carrier generation yield and the mobility.[21,54] Assuming that the latter is constant in this time window, the decay of the TRMC signal as function of time after excitation results from a decreasing concentration of mobile carriers. The sum of the effective electron and hole mobilities (i.e., the maximum TRMC signal) in the planar films is close to 30 cm^2/Vs, which is comparable to reported mobilities resulting from THz spectroscopy[21] and photoconductance studies[40,54] on similar samples. Based on the effective masses of electrons and holes in $MAPbI_3$ and the assumption that both have identical scattering times,[51,52,55] the holes are considered to be twice as mobile as the electrons. This means that the electron and hole mobilities are 10 and 20 cm^2/Vs, respectively. In contrast, for the mesostructured films, the effective mobilities are less than 6 cm^2/Vs. The lower photoconductance in mesostructured OMHPs as compared to planar films has been observed by several

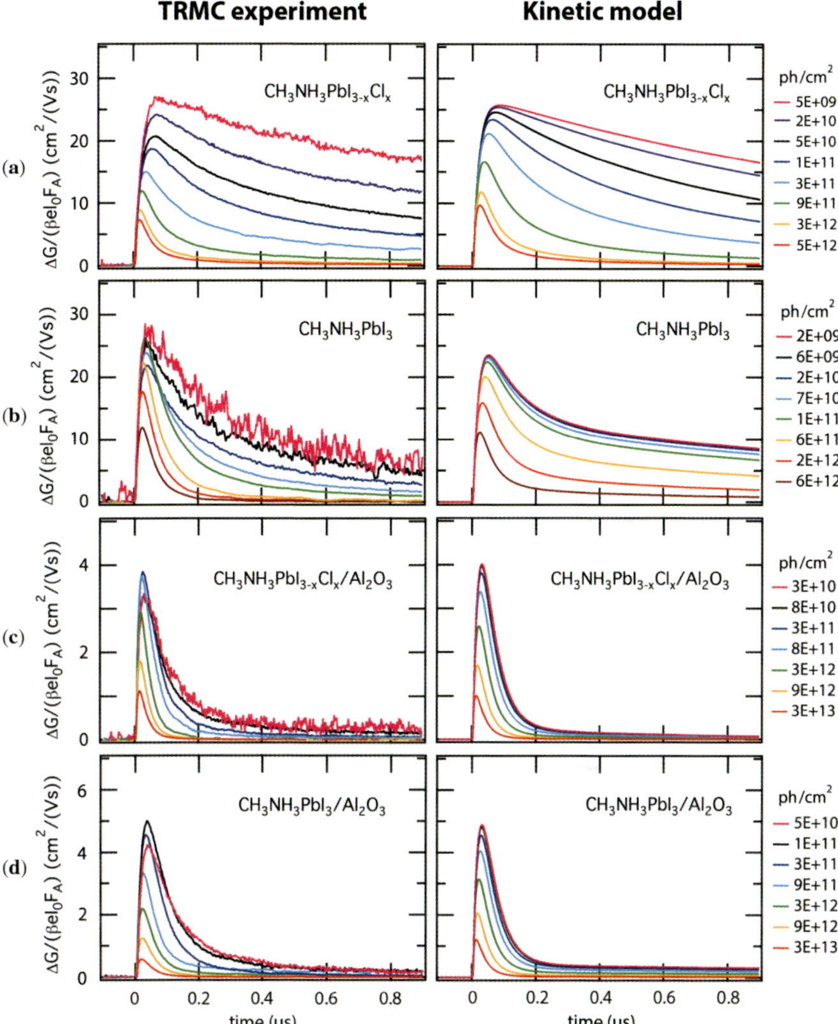

Figure 13. Experimental TRMC traces for (**a**) MAPbI$_{3-x}$Cl$_x$; (**b**) MAPbI$_3$; (**c**) MAPbI$_{3-x}$Cl$_x$/Al$_2$O$_3$; and (**d**) MAPbI$_3$/Al$_2$O$_3$ recorded at incident light intensities ranging from 10^9 to 10^{13} photons/cm^2 (λ = 600 nm, 10 Hz), corresponding to initial charge carrier densities of 10^{14}–10^{17} cm^{-3}. The right panels show the traces calculated by solving the differential eqs 2 and 3 and converting the time-dependent concentration curves to normalized ΔG. Taken from Ref. 15 with permission of the American Chemical Society.

groups.[15,56,57] This observation is most likely related to the smaller size of the crystal domains in mesostructured morphologies, which reduces the effective mobilities observed at microwave frequencies.[58–60]

At intensities below 10^{11} photons/cm^2, the lifetimes of mobile carriers are obviously substantially longer in $MAPbI_{3-x}Cl_x$ than in $MAPbI_3$ and both mesostructured films. The extended charge carrier lifetimes in $MAPbI_{3-x}Cl_x$ are particularly relevant at light intensities below 5×10^{10} photons/cm^2, because these conditions are representative for illumination by solar radiation.[61] The shorter TRMC lifetimes in $MAPbI_3$ and the mesostructured films could be attributed to immobilization of charge carriers by trapping or by recombination. For all samples, the maximum signal size lowers and the decay becomes faster with increasing laser intensity, which is indicative of the appearance of decay processes higher than first order.[13,21,49,62] That is, due to the instrumental response time of 18 ns, enhanced recombination results in a decreased intensity-normalized TRMC signal (see also Figure 5). The charge carrier mobility, on the other hand, is independent of the laser intensity.

Interestingly, this higher order recombination is most pronounced in $MAPbI_{3-x}Cl_x$, whereas $MAPbI_3$ and both mesostructured films display semi-first-order kinetics at light intensities up to 10^{11} photons/cm^2 (corresponding to $\sim 10^{15}$ excitations/cm^3). This absence of higher order decay kinetics in $MAPbI_3$ and both mesostructured films can be due to for example, trapping of electrons or recombination with dark holes (p_0, see Scheme 2). Second-order electron hole recombination will start to dominate if the concentration of photoexcited charges exceeds N_T and p_0.

Figure 14 shows semi-logarithmic time-dependent PL plots for (a) $MAPbI_{3-x}Clx$; (b) $MAPbI_3$; (c) $MAPbI_{3-x}Cl_x/Al_2O_3$; and (d) $MAPbI_3/Al_2O_3$, measured at an incident pulsed light intensity of 4×10^{12} photons/cm^2/pulse (2 J/cm^2/pulse) at 405 nm and 1 MHz. The calculated traces (black lines) using the kinetic model from Scheme 2 and fitting parameters listed in Table 1 are shown together with the experimental TRPL. The PL decay in $MAPbI_{3-x}Cl_x$ (2a) exhibits a slower decay at longer times compared to the initial decay. In contrast, the PL transient of $MAPbI_3$ (see Figure 14b, green dots) is almost linear in a semi-logarithmic plot, which corresponds to a monoexponential or pseudo-monoexponential decay.

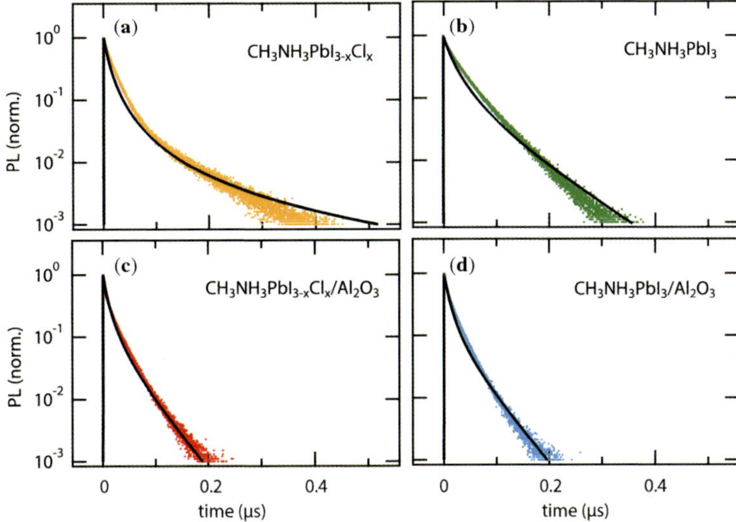

Figure 14. (a–d) PL lifetimes of photoexcited (λ = 405 nm, 1 MHz, 4 × 10^{12} photons/cm^2) thin films of OMHPs recorded at the maximum emission wavelengths shown in (Figure 12a and b). The black lines are the fits from the model. Taken from Ref. 15 with permission of the American Chemical Society.

Table 1. Overview of kinetic fitting parameters used to fit all TRMC (Figure 13) and PL (Figure 14) measurements with the model described by eqs 15–17 and summarized in Scheme 2. Taken from Ref.52 with with Permission of the American Chemical Society .

	MAPbI$_{3-x}$Cl$_x$	MAPbI$_3$	MAPbI$_{3-x}$Cl$_x$/Al$_2$O$_3$	MAPbI$_3$/Al$_2$O$_3$
Domain size (μm)	>5	± 0.5	0.05	0.05
k_2 ($x{\cdot}10^{-10}$ cm^{3}s^{-1})	4.9	3.5	7	6
k_T ($x{\cdot}10^{-10}$ cm^{3}s^{-1})	n.a.	1	3.5	2
k_R ($x{\cdot}10^{-11}$ cm^{3}s^{-1})	n.a.	5	5	1
N_T ($x{\cdot}10^{15}$ cm^{-3})	<0.5	60	7	8
p_0*($x{\cdot}10^{15}$ cm^{-3})	1	10	30	35
μ_e (cm^2/Vs)	9	10	2.5	3
μ_h (cm^2/Vs)	18	20	5	6

*Thermal equilibrium concentration at T = 300 K.

The PL lifetime is substantially shorter in the mesostructured OMHPs than in their planar analogous.[63] In addition, similar to the TRMC measurements, there is no obvious difference between the PL lifetime of $MAPbI_{3-x}Cl_x/Al_2O_3$ and $MAPbI_3/Al_2O_3$.

The nearly monoexponential decay in $MAPbI_3$ could for instance mean that the number of photogenerated electrons is very small compared to the total concentration of holes or vice versa. This could be due to background doping ($p_0 >> n_e(t)$) and/or photodoping ($n_h(t) >> n_e(t)$), the latter arising from trapping of electrons into subgap states.[13] The kinetic model accounts both for background doping and photodoping. In contrast, the PL decay of the $MAPbI_{3-x}Cl_x$ film in Figure 14a is initially second order, which indicates a regime in which electron and hole concentrations are comparable. This means that the initial concentration of photoexcited charge carriers produced by the laser pulse largely exceeds the concentration of doping levels arising from photodoping or background charges, which is consistent with the TRMC results. However, at longer time scales, the total charge density drops to reach a monoexponential decay regime, meaning that there might be a small contribution of background charges.

As mentioned before, to model the experimental TRMC and TRPL results for the planar $MAPbI_3$ film, both background carriers (p_0) and trap states (N_t) were included. This was also the case for the mesostructured films: the model describes the experimental results only if both N_T and p_0 are included. In contrast, that the trap density in the $MAPbI_{3-x}Cl_x$ thin film is too low to significantly contribute to the kinetics in the TRMC experiments. Given that measurements were performed at intensities down to 5×10^9 photons/cm^2, this means that the trap density in $MAPbI_{3-x}Cl_x$ should be less than 5×10^{14} cm^{-3}. In both thin films and mesostructured OMHPs, similar rate constants were found for band-to-band electron-hole recombination: on the order of 10^{-10} cm^{3}s^{-1}, which is comparable to earlier reports.[13,29] Furthermore, the rate constants for trap filling are only slightly lower than for band-to-band recombination. This indicates that these are competing processes if the concentration of photoinduced charges is on the same order of magnitude as the trap density. On the other hand, the rate constant for trap emptying is one order of magnitude lower, meaning that recombination between a trapped carrier and a

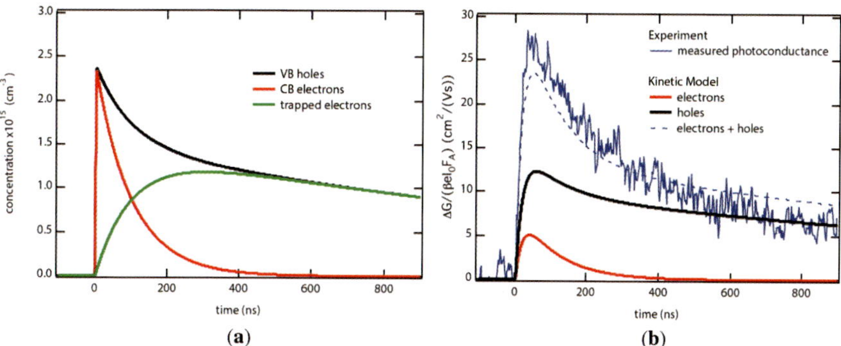

Figure 15. (**a**) Photoconductance as function of time for a 300-nm MAPbI$_3$ film (solid blue line), after excitation at 600 nm for an incident fluence of 2×10^9 photons/cm^2, together with a theoretical trace (dotted line) implementing the kinetic model from Scheme 2. Based on this model, the contributions of electrons (red) and holes (black) to the photoconductance could be separated. (**b**) Time-dependent concentration of n_e (red), n_h (black), and n_t (green) for 2.4×10^{15} photoinduced charge carriers cm^{-3}.

counExample of stop — counterconduct...

countercharge is relatively slow. This behavior results in long-lived TRMC signals originating from the mobile countercharges of the trapped carriers. The trap densities are estimated to be on the order of 10^{15} cm^{-3} in MAPbI$_{3-x}$Cl$_x$/Al$_2$O$_3$ and MAPbI$_3$/Al$_2$O$_3$ and 6×10^{16} cm^{-3} in the MAPbI$_3$ thin film. Furthermore, the model suggests a dark carrier concentration of 10^{15} cm^{-3} for MAPbI$_{3-x}$Cl$_x$ and $> 10^{16}$ cm^{-3} for the other OMHPs. Given that both mobile electrons and holes interact with the microwave field, TRMC measurements do not enable to distinguish between the two. However, after applying the kinetic model to the measured data, the time-dependent concentrations of free electrons, free holes and trapped electrons can be plotted individually. Figure 15 shows the concentration of electrons in the CB (red), trap states (green), and VB (black) as function of time after excitation of MAPbI$_3$.

The initial concentration of photoinduced electrons and holes is 2.4×10^{15} cm^{-3}. As extracted from fitting the experimental photoconductance (see Figure 15), the trap density for this sample is 6×10^{16} cm^{-3} and p_0 is 1×10^{16} cm^{-3}. Furthermore, the rate constants for electron trapping and second-order recombination are on the same order of magnitude.

Therefore, the terms $k_2 n_e (n_h + p_0)$ and $k_T n_e (N_T - n_t)$ from eqs 15 to 17 are initially comparable for these 2.4×10^{15} cm^{-3} photoinduced charge carriers. This means that electron trapping successively competes with band-to-band recombination (as shown in the right figure), which results in a shorter electron lifetime compared to the holes.

Considering that trapped electrons are immobile, the TRMC signal can be separated into the contribution from photoinduced free electrons and holes. Figure 15b shows a typical example of an experimental TRMC trace (blue solid line) together with the calculated traces for electrons (red), holes (black), and the sum of the two (blue dotted line) for the planar MAPbI$_3$ ($I_0 = 2 \times 10^9$ photons/cm^2/pulse). For this excitation density of 7×10^{13} cm^{-3}, the electron concentration has decreased to half its initial value after ~80 ns. This is defined as the half lifetime $\tau_{1/2}$. Using the electron mobility listed in Table 1, the electron diffusion length is calculated to be 1.4 μm according to:

$$L_D \sqrt{\frac{k_B T}{e} \mu \tau_{1/2}} \tag{20}$$

Similarly, the $\tau_{1/2}$ of holes is ~600 ns, which results in a hole diffusion length of 5.5 μm. Note that the electron and hole lifetimes and thus their diffusion lengths are highly dependent on the initial charge carrier concentration.

The electron and hole lifetimes obtained from applying the kinetic model to the experimental TRMC and TRPL data were used to calculate the charge-specific L_D as function of the concentration of photoinduced charges. The results are summarized in Figure 16. At 10^{15} cm^{-3} charge carriers, which is typical for steady-state AM1.5 solar illumination conditions,[61] the longest electron and hole diffusion distances are expected for planar MAPbI$_{3-x}$Cl$_x$. This mainly results from the longer lifetimes compared to the other sample types. Additionally, due to the relatively low N_T and p_0 in MAPbI$_{3-x}$Cl$_x$, the lifetimes of electrons and holes are almost the same. As a consequence, assuming that their mobilities differ not more than a factor two, the diffusion lengths of electrons and holes are very similar. On the other hand, in planar MAPbI$_3$, the diffusion length is substantially shorter for electrons than holes at excitation densities below

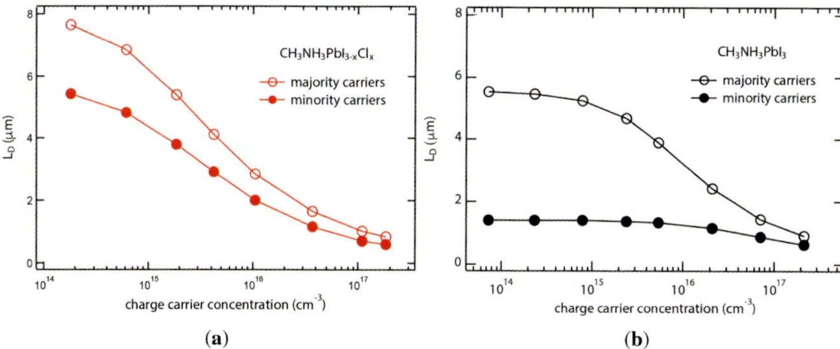

Figure 16. Diffusion length of minority carriers (closed circles) and majority carriers (open circles) as function of initial charge density for (**a**) $MAPbI_{3-x}Cl_x$ and (**b**) $MAPbI_3$, calculated by eq 20. The mobilities and lifetimes were obtained from modeling the experimental data (see Figure 13).

5×10^{16} charge carriers/cm^3. This can be explained considering that N_T and p_0 are both on the order of 10^{16} cm^{-3}, resulting in a shorter electron lifetime with respect to the holes. However, the electron diffusion length in excess of 1 μm is larger than the film thickness required to absorb all of the visible light and thus sufficient to enable efficient electron collection in a solar cell device.[64] For all samples, second-order electron-hole recombination is the dominant recombination pathway at charge carrier densities of 10^{17} cm^{-3}, resulting in analogous lifetimes for electrons and holes. If their mobilities are comparable, the electron and hole diffusion lengths will be similar. Both mesostructured films show the same TRMC and TRPL decay kinetics. Thus, it seems that using either $PbCl_2$ or PbI_2 as the precursor does not affect the electronic properties of the resulting perovskite film when this is grown within a mesoporous alumina scaffold. In contrast, the charge carrier dynamics in the $MAPbI_3$ thin film are obviously different from $MAPbI_{3-x}Cl_x$. Hence, comparing the planar thin films, it can be concluded that $PbCl_2$ instead of PbI_2 to solution-process $MAPbI_3$ substantially improves the electronic properties of the resulting film. In the kinetic model shown in Scheme 2, the OMHPs are p-type semiconductors in which the trap states are electronic in nature.[13,50,65] This purely mathematical model also holds for the opposite situation, that is, an n-type OMHP in which the trap-states are hole-selective and the electrons twice as mobile as the holes. TRMC measurements do not provide direct evidence for the type of background charges, that is, whether the films are unintentionally n-type or p-type doped.

In summary, complementary TRMC and TRPL measurements were used to investigate the charge carrier dynamics in planar and mesoporous perovskite films prepared with different precursors. A kinetic model was applied to the experimental data to calculate the concentrations of background charges and trap states. The observation that global fitting could be performed to both TRMC and TRPL measurements confirms that recombination of free mobile charges gives rise to PL in $MAPbI_3$ thin films.

B. Single Crystal Perovskite

Millimeter-sized single OMHP crystals are an ideal model system to investigate intrinsic material properties, such as the charge carrier recombination pathways in the absence of grain boundaries.[66–70] TRMC, TRPL, and the kinetic model from Scheme 2 were used to analyze $MAPbI_3$ single crystals with dimensions of 10 mm by 8 mm by 3 mm.[71] These were grown in a supersaturated $MAPbI_3$ precursor solution using the top-seeded-solution-growth method.[72] Different excitation wavelengths were used to manipulate the location of charge carrier generation and study the effect on the recombination dynamics. As shown in Figure 17, the penetration depth $(1/\alpha)$ of visible light in $MAPbI_3$ is typically a few hundred nanometers: at least three orders of magnitude smaller than the total depth of the single crystal. Therefore, visible light can be used to study the dynamics of charge carriers relatively close to the surface. On the other hand, an initially homogeneous charge carrier concentration can only be realized if α is substantially lower, which is the case for excitation wavelengths close to the absorption onset. An integrating sphere was used to collect both specular and diffuse reflectance of a single $MAPbI_3$ crystal (Figure 17a). By subtracting these from the incident light intensity, the absorption spectrum of the single crystal was obtained. This is displayed in Figure 17b, where α is plotted in the inset for excitation wavelengths between 820 and 890 nm.

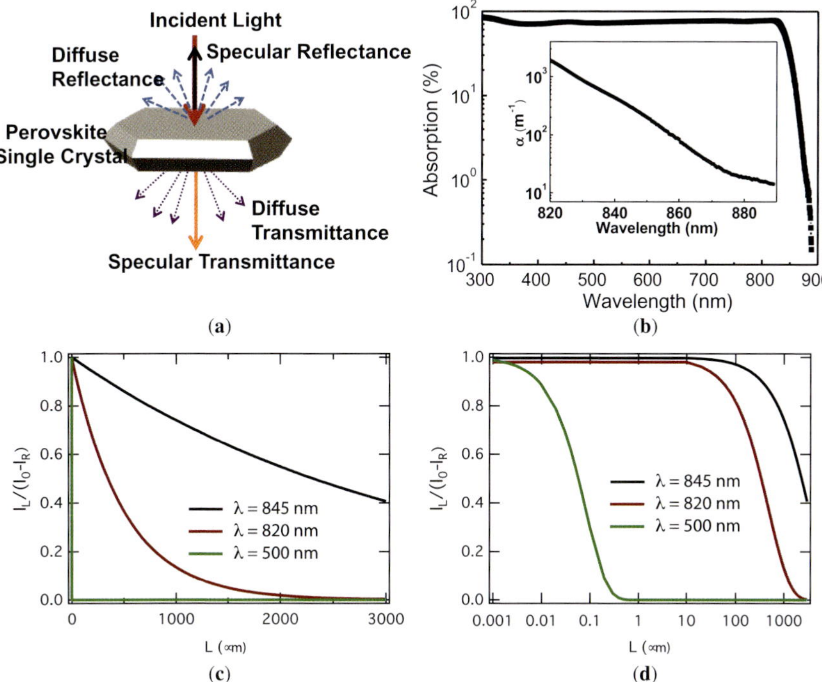

Figure 17. (a) Overview of absorption measurement on a MAPbI$_3$ single crystal. (b) Light absorption (%) and absorption coefficient (m^{-1}) close to the offset of the band gap. The absorption coefficients at 820 nm (2×10^3 m^{-1}) and 845 nm (3×10^2 m^{-1}) were used to model the excitation profile; (c and d) where the x-axis corresponds to the total thickness of the single crystal. The excitation profile at 500 nm was calculated using the absorption coefficient from Figure 1a. Taken from Ref. 71 with permission of the American Chemical Society.

The wavelength-dependent absorption coefficients from Figures 1a and 17b were used to calculate the charge carrier generation profile as function of the excitation wavelength. The results are shown on a linear scale in Figure 17c for 500, 820, and 845 nm and a semi-logarithmic scale in Figure 17d. This visualizes the extremely different distribution of charge carriers in a 3-mm thick crystal for these excitation wavelengths. With 845 nm light, which is close to the absorption onset as shown in Figure 17b, a homogeneous concentration profile can be realized because $1/\alpha$ is 0.33 cm and thus close to the thickness of the crystal.

The MAPbI$_3$ single crystal was glued on a quartz substrate and mounted in the microwave cell to perform the TRMC measurements. Given that the single crystal is extremely large with respect to thin films, the sensitivity factor K of the microwave cell can no longer be calculated as described earlier. Without accurate determination of K, ΔG, and thus $\varphi\Sigma\mu$ cannot be calculated from $\Delta P/P$. Therefore, $\Delta P/P$ was only corrected for I_0 to compare the different excitation wavelengths and photon fluences.

Figure 18 shows $-\Delta P/PI_0$ of the MAPbI$_3$ crystal as function of time for excitation wavelengths of 845 and 500 nm (Figure 18a and b,

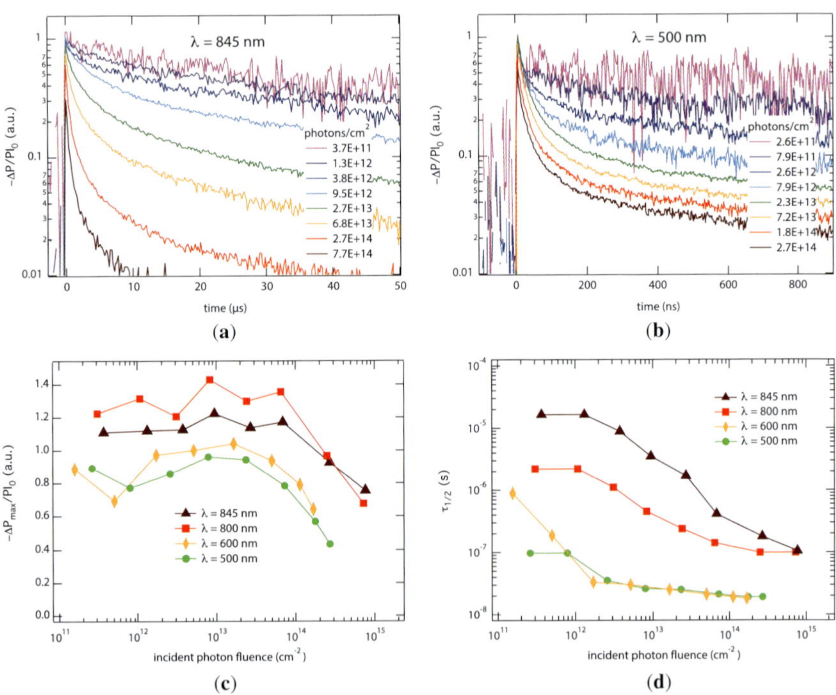

Figure 18. TRMC results under different excitation wavelengths and intensities. TRMC traces normalized to the incident intensity for the perovskite crystal recorded at excitation wavelength (**a**) 845 nm and (**b**) 500 nm using different indicated excitation intensities (photons/cm^2). Note that the horizontal scales are different. (**c**) Maximum change in normalized microwave power corrected for the amount of absorbed photons and (**d**) corresponding half lifetimes as function of the photon fluence at indicated wavelengths. Taken from Ref. 71 with permission of the American Chemical Society.

respectively). Figure 18c shows the maximum signal sizes as function of the photon fluence at different excitation wavelengths (500, 600, 800, and 845 nm), showing somewhat smaller values for the visible wavelengths (500 and 600 nm). In Figure 18d, $\tau_{1/2}$ values are plotted for incident intensities ranging from 10^{11} to 10^{15} photons/cm^2. On excitation at 845 nm, values of $\tau_{1/2}$ over 15 μs are found at low intensities ($<10^{12}$ photons per cm^2, $\sim10^{12}$ charges/cm^3). However, upon increasing the intensity the $\tau_{1/2}$ reduces down to about 100 ns. For excitation at 500 and 600 nm, $\tau_{1/2}$ values show a reduction from a few hundreds of nanoseconds down to about 15 ns on increasing intensities, which is indicate for higher order recombination pathways.[15,16,21,71] The reduction of the maximum signal sizes on higher intensities in Figure 18c could thus be the result from increased recombination within the response time of the measurements. Upon close inspection of the TRMC traces recorded on excitation at 500 nm (Figure 18b), it is noticed that in particular at higher intensities a small additional peak can be discerned within the first tens of nanoseconds, which is absent for excitation at 845 nm. This strong decay can be explained in two ways: (1) fast second-order recombination due to the high local photoinduced charge carrier concentration (10^{17} cm^{-3}) and (ii) in addition, strong charge recombination at the surface of the single crystals might occur. If the surface of the crystal acts as an efficient recombination site, one would expect a smaller signal size on optical excitation at 500 and 600 nm. This is in agreement with the observations shown in Figure 18. For these wavelengths, the largest part of the charge carriers is generated within 100 nm of the surface. Assuming a modest charge carrier mobility of 10 cm^2/Vs these carriers have definitely encountered the crystal surface within a few nanoseconds. The absence of this fast initial decay at 800 and 845 nm is consistent with the fact that, at these wavelengths, most photoinduced carriers are generated further away from the surface.

To further investigate the charge recombination channels in the single crystals, the kinetic model from Scheme 2 was applied. The initial charge carrier concentration was calculated using the penetration depth instead of the crystal thickness (see eq 2) and the absorption coefficient shown in Figure 1. The results are shown in Figure 19 for the TRMC decay kinetics obtained at 845 nm and the parameters used to fit all the different intensities are listed in Table 2. An N_T of 1.5×10^{13} cm^{-3} was found, which is at

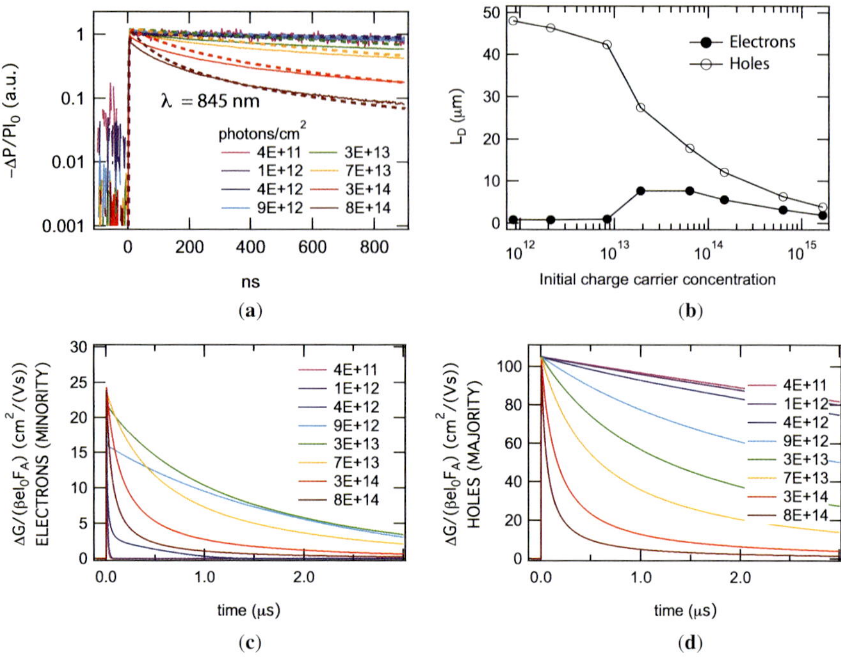

Figure 19. (**a**) Fits (dashed lines) to the TRMC traces (full lines) using the model as described in the text for excitation at 845 nm, where charges are generated homogeneously throughout the crystal. (**b**) Semi-logarithmic plot of the diffusion length for majority carriers (holes, blue) and minority carriers (electrons, yellow) as function of the concentration of photoexcited charge carriers. (**c**) CB electrons and (**d**) VB holes (photons/cm^2, λ = 845 nm). Note that the experimental TRMC traces are always the sum of (c) and (d). Taken from Ref. 71 with permission of the American Chemical Society.

least one order of magnitude lower than values found in the thin MAPbI$_3$ films (see also Table 1).[6,11,21,29] This discrepancy could be attributed to the lack of grain boundaries within the single crystals. Furthermore, the k_2 is 5.5×10^{-9} cm^{3}s^{-1}, which is about one order of magnitude higher than reported in several other studies.[15,13,29,49] The higher charge carrier mobility caused by the superior crystallinity in the crystal could result in an enhancement of k_2.

Similar to the analysis performed on the thin films, after fitting the TRMC traces with the kinetic model, the electron and hole contributions to the photoconductance were separated. The results are shown in

Table 2. Kinetic parameters used to model the TRMC measurements (see Figure 19a). Taken from Ref. 71 with permission of the American Chemical Society.

k_2 (cm^{3}s^{-1})	5.5×10^{-9}
k_T (cm^{3}s^{-1})	9×10^{-6}
k_R (cm^{3}s^{-1})	2×10^{-9}
N_T (cm^{-3})	1.5×10^{13}
p_0 (cm^{-3})	4×10^{13}
$\Sigma\mu_h$ *(cm^2/Vs)	105
$\Sigma\mu_e$ *(cm^2/Vs)	25

Here, k_2, k_T, and k_R are the rate constants for band-to-band electron-hole recombination, trap filling, and trap emptying, respectively. N_T denotes the concentration of trap states, p_0 is the background hole concentration at thermal equilibrium. Finally, μ_e and μ_h are the mobilities of electrons (e) and holes (h).

Figure 19b, assuming electron and hole mobilities of 25 and 105 cm^2/Vs, respectively, as reported for this type of single crystals.[67] Figure 19c and d shows that at light intensities below 4×10^{12} photons/cm^2, the hole lifetime (microseconds) is much longer than the electron lifetime (<50 ns). On the other hand, when the concentration of photoexcited electrons exceeds the number of trap states (e.g., at 3×10^{13} photons/cm^2), the electron lifetime is also in the order of microseconds.

Figure 19b shows the electron and hole diffusion lengths as function of the initial concentration (calculated from eq 20) of photogenerated electron-hole pairs. These results show that below the trap concentration (<1.5×10^{13} cm^{-3}), the electron diffusion length is 0.8 μm, which increases to 8 μm just above the trap concentration. For the holes, the diffusion length reaches a maximum value close to 50 μm, which is attributed to the slow recombination with a trapped electron.[13,15] At higher concentrations, the diffusion lengths of both electrons and holes reduce due to increasing second-order electron-hole recombination. These results suggest that the diffusion lengths will be optimal at an electron-hole concentration just above the trap density. Note that this model applies to a homogeneous,

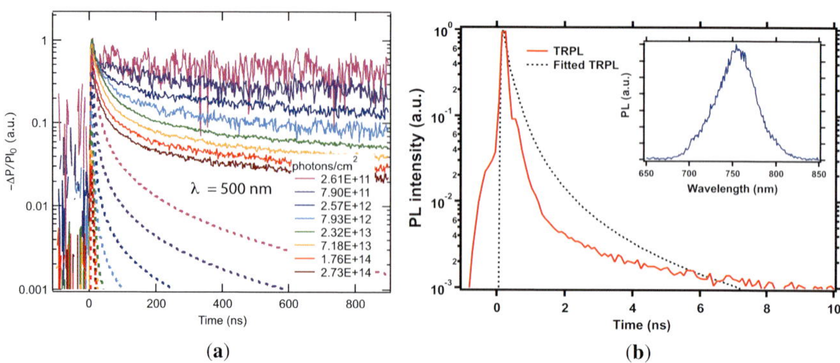

Figure 20. (**a**) Modeled TRMC traces (dashed lines) using the kinetic parameters obtained from fitting the data shown in Figure 19a, see Table 2, together with the experimental TRMC traces (full lines) recorded at an excitation wavelength of 500 nm (see also Figure 18). (**b**) TRPL trace of single crystal perovskite on pulsed excitation at 405 nm with an intensity 4×10^{12} photons cm^{-2}, detected at 760 nm (dashed line) and PL trace calculated with the model described in the text (full line). The inset shows the emission spectrum of the single crystal MAPbI$_3$ using an excitation wavelength of 405 nm. Taken from Ref. 71 with permission of the American Chemical Society.

pulsed excitation profile: additional diffusion resulting from a concentration gradient is not taken into account.

The kinetic parameters obtained from fitting the TRMC decay kinetics at $\lambda = 845$ nm were used to calculate the TRMC traces for excitation at $\lambda = 500$ nm, using an initial concentration based on a charge carrier generation depth of $x = 85$ nm. The fitted traces are plotted together with the experimental TRMC data in Figure 20. Considering that initially, the local concentration of electrons and holes is extremely high for excitation at $\lambda = 500$ nm, the calculated lifetimes are substantially shorter than for $\lambda = 845$ nm. However, as shown in Figure 20a, the calculated traces decay much faster than those observed experimentally at $\lambda = 500$ nm. This means that recombination is slower than what would be expected based on the locally high concentration. This suggests that electron and hole concentrations are lower than calculated assuming a generation depth of 85 nm. Most likely, due to the high initial charge carrier concentration gradient, carriers have diffused toward the bulk of the crystal leading to lowered concentrations and hence longer lifetimes. Therefore, an accurate description of the charge carrier dynamics in single crystals for $\lambda < 800$ nm requires an extension of the kinetic model, accounting for diffusion. Finally, PL measurements were performed on the same single crystal (λ = 405 nm, 4×10^{12} photons/cm^2). Figure 20 shows the TRPL with the emission spectrum in the inset. Comparable to previously published values,[73] the PL lifetime is a few nanoseconds and hence, much shorter than observed in thin film (see also Figure 14). Additionally, the PL decay is substantially faster than the TRMC decays: under similar excitation conditions (i.e., λ = 500 nm, 2.6×10^{12} photons/cm^2) the lifetime of free mobile charges is many microseconds as shown in Figure 20. Using the parameters listed in Table 2, found for fitting the TRMC traces at 845 nm, and a generation depth of x = 12.5 nm (i.e., $1/2\alpha$) at 405 nm, a TRPL trace is calculated as described earlier and added in Figure 20. The measured PL decay is close to the calculated trace. Considering that the charge carrier concentration was estimated using $1/2\alpha$, the model only accounts for the emission of the surface layer with a thickness of half the penetration depth. This yields a very high surface concentration of charge carriers, which undergo rapid band-to-band recombination leading to the fast decay kinetics. PL emitted more in the interior of the crystal is likely to be reabsorbed and will therefore not be probed.[73] These results show that the short PL lifetimes, as typically observed for single crystals can be explained by the relatively high recombination rate k_2 in combination with the high local concentration of charge carriers close to the surface of the crystal that recombine radiatively. Another part of the charge carriers diffuses into the crystal, resulting in microsecond lifetimes of free, mobile charges as detected with the TRMC. Hence, the differences observed in the decay kinetics for the single crystal using PL and TRMC highlight the dissimilarities in measuring techniques. While PL detects only radiative recombination close to the surface, TRMC probes all mobile carriers formed within the crystal. From this, it can be deduced that for single crystals, the PL lifetimes are not representative of the lifetime of free mobile charges within the crystal and, therefore, not suitable to determine the charge carrier diffusion lengths. Instead of $\tau_{1/2}$ values of several nanoseconds as found by TRPL, TRMC reveals that under low excitation energies (<10^{12} photons per cm^2), $\tau_{1/2}$ can be as large as 15 μs.

In summary, TRMC measurements were performed on a millimeter-sized single MAPbI$_3$ crystal, using different excitation wavelengths to manipulate the initial location of photogenerated charge carriers. For excitation at 845 nm, a homogeneous distribution of charges was realized. From here, using the same approach as for the thin films, the bulk trap density in a single $MAPbI_3$ was determined to be on the order of 10^{13} cm^{-3}. Additionally, these results indicate that the majority and minority carrier diffusion lengths in $MAPbI_3$ can be as large as 50 and 10 μm, respectively, if it is no longer limited by the dimensions of the crystallites.

V. Conclusions and Perspectives

The evolution of the charge carrier dynamics in perovskite solar cell materials was monitored by obtaining the transient photoconductivity in the time scales of sub-picosecond to few microseconds. In the early time scale, the initial photoproduct was identified as mostly highly mobile charges with some contribution coming from loosely bound exciton whose ratio may be different depending on the homogeneity of the quality of the film. These photogenerated charges maintain its high mobility and do not recombine in the nanosecond time scale. Moreover, the time scale and injection mechanism of electrons and holes, to either organic or metal oxide electrodes, are dictated by the alignment of the band energy levels at the interface of the materials. The extraction of injected charges at the interface is inhibited by the low conductivity of the acceptor materials inducing recombination with the charges in perovskite. On the nanosecond to microsecond time scale, free charges recombine *via* second-order recombination, which is the origin of PL. If the charge carrier concentration is lower than the concentration of trap states, radiative recombination is in competition with trapping of electrons or holes. Finally, analysis on millimeter-sized single crystals show that diffusion lengths in $MAPbI_3$ could be tens of micrometers if charge carrier transport is not limited by the size of the crystalline domains.

VI. Acknowledgment

This work was supported by the Swedish Energy Agency (STEM), the Swedish Research Council, the Knut and Alice Wallenberg foundation, the nanometer Consortium at Lund University (nmc@LU), the Lund Laser Center, and the Laserlab-Europe (EU-H2020 654148), and the

Netherlands Organization for Scientific Research (NWO) under the Echo grant number: 712.014.007.

VII. References

1. Kojima, A.; Teshima, K.; Shirai, Y.; Miyasaka, T. *J. Am. Chem. Soc.* **2009**, *48*, 6050.

2. National Renewable Energy Labs (NREL) Efficiency Chart 2016. http://www.nrel.gov/ncpv/images/efficiency_chart.jpg Natl Renew Energy Labs Effic chart. 2016.

3. Tan, Z. K.; Moghaddam, R. S.; Lai, M. L.; Docampo, P.; Higler, R.; Deschler, F.; Price, M.; Sadhanala, A.; Pazos, L. M.; Credgington, D.; Hanusch, F.; Bein, T.; Snaith, H. J.; Friend, R. H. *Nat. Nanotechnol.* **2014**, *8*, 737–743.

4. Zhu, H.; Fu, Y.; Meng, F.; Wu, X.; Gong, Z.; Ding, Q.; Gustafsson, M. V.; Trinh, M. T.; Jin, S.; Zhu, X.-Y. *Nat. Mater.* **2015**, *14*, 636–642.

5. Luo, J.; Im, J.-H.; Mayer, M. T.; Schreier, M.; Nazeeruddin, M. K.; Park, N.-G.; Tilley, S. D.; Fan, H. J.; Gratzel, M. *Science* **2014**, *345*, 1593–1596.

6. Baxter, J. B.; Schmuttenmaer, C. A. *J. Phys. Chem. B* **2006**, *110*, 25229–25239.

7. Ponseca, C. Jr. S.; Sundstrom, V. *Nanoscale* **2016**, *8*, 6249–6257.

8. Kroeze, J. E.; Savenije, T. J.; Vermeulen, M. J. W.; Warman, J. M. *J. Phys. Chem. B* **2003**, *107*, 7696–7705.

9. Sandeep, C. S. S.; ten Cate, S.; Schins, J. M.; Savenije, T. J.; Liu, Y.; Law, M.; Kinge, S.; Houtepen, A. J.; Siebbeles, L. D. a. *Nat. Commun.* **2013**, *4*, 2360(1–4).

10. Savenije, T. J.; Ferguson, A. J.; Kopidakis, N.; Rumbles, G. *J. Phys. Chem. C* **2013**, *117*, 24085–24103.

11. Dicker, G.; De Haas, M. P.; Siebbeles, L. D. A.; Warman, J. M. *Phys. Rev. B* **2004**, *70*, 045203.

12. Murthy, D. H. K.; Melianas, A.; Tang, Z.; Juška, G.; Arlauskas, K.; Zhang, F.; Siebbeles, L. D. A.; Inganäs, O.; Savenije, T. J. *Adv. Funct. Mater.* **2013**, *23*, 4262–4268.

13. Stranks, S. D.; Burlakov, V. M.; Leijtens, T.; Ball, J. M.; Goriely, A.; Snaith, H. J. *Phys. Rev. Appl.* **2014**, *2*, 034007(1–4).

14. Miyata, A.; Mitioglu, A.; Plochocka, P.; Portugall, O.; Wang, J. T.-W.; Stranks, S. D.; Snaith, H. J.; Nicholas, R. J. *Nat. Phys.* **2015**, *11*, 582–587.

15. Hutter, E. M.; Eperon, G. E.; Stranks, S. D.; Savenije, T. J. *J. Phys. Chem. Lett.* **2015**, *6*, 3082–3090.

16. Savenije, T. J.; Ponseca, Carlito S, J.; Kunneman, L.; Abdellah, M.; Zheng, K.; Tian, Y.; Zhu, Q.; Canton, S. E.; Scheblykin, I. G.; Pullerits, T.; Yartsev, A.; Sundström, V. *J. Phys. Chem. Lett.* **2014**, *5*, 2189–2194.

17. Lin, Q.; Armin, A.; Nagiri, R. R. C. R.; Burn, P. L.; Meredith, P.; Chandra, R. *Nat. Photonics* **2014**, *9*, 106–112.

18. Sheng, C.; Zhang, C.; Zhai, Y.; Mielczarek, K.; Wang, W.; Ma, W.; Zakhidov, A.; Vardeny, Z. *Phys. Rev. Lett.* **2014**, *114*, 116601.

19. Ne☐mec, H.; Rochford, J.; Taratula, O.; Galoppini, E.; Kužel, P.; Polívka, T.; Yartsev, A.; Sundström, V. *Phys. Rev. Lett.* **2010**, *104*, 197401.

20. Zidek, K.; Zheng, K.; Ponseca, C. S., Jr.; Messing, M. E.; Wallenberg, L. R.; Chabera, P.; Abdellah, M.; Sundström, V.; Pullerits, T. *J. Am. Chem. Soc.* **2012**, *134*, 12110–12117.

21. Ponseca, C. S.; Savenije, T. J.; Abdellah, M.; Zheng, K.; Yartsev, A.; Pascher, T.; Harlang, T.; Chabera, P.; Pullerits, T.; Stepanov, A.; Wolf, J.-P.; Sundström, V. *J. Am. Chem. Soc.* **2014**, *136*, 5189–5192.

22. Hendry, E.; Koeberg, M.; O'Regan, B.; Bonn, M. *Nano Lett.* **2006**, *6*, 755–759.

23. Ponseca, C. S., Jr.; Ne☐mec, H.; Vukmirović, N.; Fusco, S.; Wang, E.; Andersson, M.; Chavera, P.; Yartsev, A.; Sundström, V. *J. Phys. Chem. Lett.* **2012**, *3*, 2442–2446.

24. Blom, P. W. M.; Mihailetchi, V. D.; Koster, L. J. A.; Markov, D. E. *Adv. Mater.* **2007**, *19*, 1551.

25. Umari P.; Mosconi, E.; Angelis, F. D. *Sci. Rep.* **2013**, *4*, 4467.

26. Ponseca, C. S., Jr.; Yartsev, A.; Wang, E.; Andersson, M.; Vithanage, D.; Sundström, V. *J. Am. Chem. Soc.* **2012**, *134*, 11836–11839.

27. Leguy, A. M. A.; Frost, J. M.; McMahon, A. P.; Sakai, V. G.; Kochelmann, W.; Law, C.; Li, X.; Foglia, F.; Walsh, A.; O'Regan, B. C.; Nelson, J.; Cabral, J. T.; Barnes, P. R. *Nature Comms.* **2015**, *6*, 7124.

28. Yang, W. S.; Noh, J. H.; Jeon, N. J.; Kim, Y. C.; Ryu, S.; Seo, J.; Seok, S. I. *Science* **2015**, *348*, 1234–1237.

29. Eperon, G. E.; Stranks, S. D.; Menelaou, C.; Johnston, M. B.; Herz, L. M.; Snaith, H. J. *Energy Environ.l Sci.* **2014**, *7*, 982–988.

30. Rehman, W.; Milot, R. L.; Eperon, G. E.; Wehrenfennig, C.; Boland, J. L.; Snaith, H. J.; Johnston, M. B.; Herz, L. M. *Adv. Mater.* **2015**, *27*, 7938–7944.

31. Kieslich, G.; Sun, S.; Cheetham, A. K. *Chem. Sci.* **2014**, *5*, 4712–4715.

32. Piatkowski, P.; Cohen, B.; Ponseca, C. S., Jr.; Salado, M.; Kazim, S.; Ahmad, S.; Sundstrom, V.; Douhal, A. *J. Phys. Chem. Lett.* **2016**, *7*, 204–210.

33. Nemec, H.; Kuzel, P.; Sundstrom, V. *J. Photochem. Photobiol. A* **2010**, *215*, 123–230.

34. Ulbricht, R.; Hendry, E.; Shan, J.; Heinz, T. F.; Bonn, M. *Rev. Mod. Phys.* **2011**, *83*, 543–586.

35. Park, J. H.; Seo, J.; Park, S.; Shin, S. S.; Kim, Y. C.; Jeon, N. J.; Shin, H. W.; Ahn, T. K.; Noh, J. H.; Yoon, S. C.; Hwang, C. S.; Seok, S. I. *Adv. Mater.* **2015**, *27*, 4013–4019.

36. Che, W.; Wu, Y.; Yue, Y.; Liu, J.; Zhang, W.; Yang, X.; Chen, H.; Bi, E.; Ashraful, I.; Graetzel, M.; Han, L. *Science* **2015**, *350*, 944–948.

37. You, J.; Meng, L. M.; Song, B. S.; Guo, T.-F.; Yang, M. Y.; Chang, W.-H.; Hong, Z.; Chen, H.; Zhou, H.; Chen, Q.; Liu, Y.; De Marco, N.; Yang, Y. *Nat. Nanotechnol.* **2016**, *11*, 75–81.

38. Wang, Q.; Shao, Y.; Xie, H.; Lyu, L.; Liu, X.; Gao, Y.; Huang, J. *Appl. Phys. Lett.* **2014**, *105*, 163508.

39. Tian, Y.; Scheblykin, I. G. *J. Phys. Chem. Lett.* **2015**, *6*, 3466–3470.

40. Wittenauer, M. A.; Van Zandt, L. L. S *Philos. Mag. B* **1982**, *46*, 659–667.

41. Corani, A.; Li, M.H.; Shen, P.-S.; Guo, T. F.; El Nahhas, A.; Zheng, K.; Yartsev, A.; Sundström, V.; Ponseca, C. Jr. S. *J. Phys. Chem. Lett.* **2016**, *7*, 1096–1101.

42. Ponseca, C. Jr. S.; Hutter, E.M.; Piatkowski, P.; Cohen, B.; Pascher, T.; Douhal, A.; Yartsev, A.; Sundstrom, V.; Savenije, T. J. *J. Am. Chem. Soc.* **2015**, *137*, 16043–16048.

43. Abrusci, A.; Stranks, S. D.; Docampo, P.; Yip, H.-L.; Jen, A. K.-Y.; Snaith, H. J. *Nano Lett.* **2013**, *13*, 3124–3128.

44. Nguyen, W. H.; Bailie, C. D.; Unger, E. L.; McGehee, M. D. *J. Am. Chem. Soc.* **2014**, *136*, 10996–11001.

45. Xing, G.; Mathews, N.; Sun, S.; Lim, S. S.; Lam, Y. M.; Grätzel, M.; Mhaisalkar, S.; Sum, T. C. *Science* **2013**, *342*, 344–347.

46. Eperon, G. E.; Burlakov, V. M.; Docampo, P.; Goriely, A.; Snaith, H. J. *Adv. Funct. Mater.* **2014**, *24*, 151–157.

47. Stranks, S. D.; Nayak, P. K.; Zhang, W.; Stergiopoulos, T.; Snaith, H. J. *Angew. Chemie Int. Ed.* **2015**, *54*, 3240–3248.

48. Lee, M. M.; Teuscher, J.; Miyasaka, T.; Murakami, T. N.; Snaith, H. J. *Science* **2012**, *338*, 643–647.

49. Wehrenfennig, C.; Eperon, G. E.; Johnston, M. B.; Snaith, H. J.; Herz, L. M. *Adv. Mater.* **2014**, *26*, 1584–1589.

50. Leijtens, T.; Stranks, S. D.; Eperon, G. E.; Lindblad, R.; Johansson, E. M. J.; Mcpherson, I. J.; Rensmo, H.; Ball, J. M.; Lee, M. M.; Snaith, H. J. *ACS Nano* **2014**, *8*, 7147–7155.

51. Chang, Y.; Park, C.; Matsuishi, K. *J. Korean Phys. Soc.* **2004**, *44*, 889–893.

52. Giorgi, G.; Fujisawa, J. I.; Segawa, H.; Yamashita, K. *J. Phys. Chem. Lett.* **2013**, *4*, 4213–4216.

53. Holzwarth, U.; Gibson, N. *Nat. Nanotechnol.* **2011**, *6*, 534.

54. Wojciechowski, K.; Stranks, S. D.; Abate, A.; Sadoughi, G.; Sadhanala, A.; Kopidakis, N.; Rumbles, G.; Li, C.-Z.; Friend, R. H.; Jen, A. K.-Y.; Snaith, H. J. *ACS Nano* **2014**, *8*, 12701–12709.

55. Quarti, C.; Mosconi, E.; De Angelis, F. *Chem. Mater.* **2014**, *26*, 6557–6569.

56. Pascoe, A. R.; Yang, M.; Kopidakis, N.; Zhu, K.; Reese, M. O.; Rumbles, G.; Fekete, M.; Duffy, N. W.; Cheng, Y.-B. *Nano Energy* **2016**, *22*, 439–452.

57. Oga, H.; Saeki, A.; Ogomi, Y.; Hayase, S.; Seki, S. *J. Am. Chem. Soc.* **2014**, *136*, 13818–13825.

58. Wang, Y. C.; Ahn, H.; Chuang, C. H.; Ku, Y. P.; Pan, C. L. *Appl. Phys. B* **2009**, *97*, 181.

59. Katoh, R.; Huijser, A.; Hara, K.; Savenije, T. J.; Siebbeles, L. D. A. *J. Phys. Chem. C* **2007**, *111*, 10741–10746.

60. Prins, P.; Grozema, F. C.; Siebbeles, L. D. A. *J. Phys. Chem. B* **2006**, *110*, 14659–14666.

61. Johnston, M. B.; Herz, L. M. *Acc. Chem. Res.* **2016**, *49*, 146–154.

62. Manser, J. S.; Kamat, P. V. *Nat. Photonics* **2014**, *8*, 737–743.

63. De Bastiani, M.; D'Innocenzo, V.; Stranks, S. D.; Snaith, H. J.; Petrozza, A.; Bastiani, M. De; Innocenzo, V. D. *APL Mater.* **2014**, *2*, 081509.

64. Stranks, S. D.; Eperon, G. E.; Grancini, G.; Menelaou, C.; Alcocer, M. J. P.; Leijtens, T.; Herz, L. M.; Petrozza, A.; Snaith, H. J. *Science* **2013**, *342*, 341–344.

65. Wetzelaer, G.-J. A. H.; Scheepers, M.; Sempere, A. M.; Momblona, C.; Avila, J.; Bolink, H. J. *Adv. Mater.* **2015**, *27*, 1837–1841.

66. Dang, Y.; Liu, Y.; Sun, Y.; Yuan, D.; Liu, X.; Lu, W.; Liu, G.; Xia, H.; Tao, X. *CrystEngComm* **2015**, *17*, 665–670.

67. Dong, Q.; Fang, Y.; Shao, Y.; Mulligan, P.; Qiu, J.; Cao, L.; Huang, J. *Science* **2015**, *347*, 967–970.

68. Saidaminov, M. I.; Abdelhady, A. L.; Murali, B.; Alarousu, E.; Burlakov, V. M.; Peng, W.; Dursun, I.; Wang, L.; He, Y.; Maculan, G.; Goriely, A.; Wu, T.; Mohammed, O. F.; Bakr, O. M. *Nat. Commun.* **2015**, *6*, 7586.

69. Fang, Y.; Dong, Q.; Shao, Y.; Yuan, Y.; Huang, J. *Nat. Photonics* **2015**, *9*, 679–686.

70. Shi, D.; Adinolfi, V.; Comin, R.; Yuan, M.; Alarousu, E.; Buin, A.; Chen, Y.; Hoogland, S.; Rothenberger, A.; Katsiev, K.; Losovyj, Y.; Zhang, X.; Dowben, P. A.; Mohammed, O. F.; Sargent, E. H.; Bakr, O. M. *Science* **2015**, *347*, 519–522.

71. Bi, Y.; Hutter, E. M.; Fang, Y.; Dong, Q.; Huang, J.; Savenije, T. J. *J. Phys. Chem. Lett.* **2016**, *7*, 923–928.

72. de Quilettes, D. W.; Vorpahl, S. M.; Stranks, S. D.; Nagaoka, H.; Eperon, G. E.; Ziffer, M. E.; Snaith, H. J.; Ginger, D. S. *Science* **2015**, *348*, 683–686.

73. Yamada, Y.; Yamada, T.; Phuong, L. Q.; Maruyama, N.; Nishimura, H.; Wakamiya, A.; Murata, Y.; Kanemitsu, Y. *J. Am. Chem. Soc.* **2015**, *137*, 10456–10459.

Index